IEE Telecommunications Series 19
Series Editors: Professors J. E. Flood, C. J. Hughes and J. D. Parsons

Telecommunications Traffic, Tariffs and Costs

An Introduction for Managers

Other volumes in this series

Volume 1   Telecommunications networks
           J. E. Flood (Editor)
Volume 2   Principles of telecommunication-traffic engineering
           D. Bear
Volume 3   Programming electronic switching systems
           M. T. Hills and S. Kano
Volume 4   Digital transmission systems
           P. Bylanski and D. G. W. Ingram
Volume 5   Angle modulation: the theory of system assessment
           J. H. Roberts
Volume 6   Signalling in telecommunications networks
           S. Welch
Volume 7   Elements of telecommunications economics
           S. C. Littlechild
Volume 8   Software design for electronic switching systems
           S. Takamura, H. Kawashima, N. Nakajima
Volume 9   Phase noise in signal sources
           W. P. Robins
Volume 10 Local telecommunications
           J. M. Griffiths (Editor)
Volume 11 Principles and practice of multi-frequency telegraphy
           J. D. Ralphs
Volume 12 Spread spectrum in communications
           R. Skaug and J. F. Hjelmstad
Volume 13 Advanced signal processing
           D. J. Creasey (Editor)
Volume 14 Land mobile radio systems
           R. J. Holbeche (Editor)
Volume 15 Radio receivers
           W. Gosling (Editor)
Volume 16 Data communications and networks
           R. L. Brewster (Editor)
Volume 17 Local telecommunications 2
           J. M. Griffiths (Editor)
Volume 18 Satellite communication systems
           B. G. Evans (Editor)

Telecommunications Traffic, Tariffs and Costs

An Introduction for Managers

R E Farr

Peter Peregrinus Ltd on behalf of the Institution of Electrical Engineers

Published by: Peter Peregrinus Ltd, London, United Kingdom

© 1988 Peter Peregrinus Ltd

ISBN 0 86341 145 2

Printed in England by Short Run Press Ltd, Exeter

Contents

		page
Acknowledgment		
Introduction		
✔ Chapter 1	The nature of telecommunications traffic	1
1.1	Telephone traffic	1
1.2	The effect of unsuccessful calls on telephone traffic flow	3
1.3	Traffic-flow units and circuit occupancy	5
1.4	Loss or delay working and repeat attempts	5
1.5	Variability of traffic flow: the busy hour	7
1.6	Grade of service	8
1.7	Data traffic	9
1.8	The store-and-forward concept and delay	10
1.9	Packet-type data traffic	12
Chapter 2	The transmission of telephone traffic	14
2.1	Signalling and analogue transmission over the local network	14
2.2	Beyond the local network: long-distance analogue transmission	15
2.3	The need for international standards	18
2.4	Beyond the local network: digital transmission	19
2.5	Digital transmission over optical fibres	23
2.6	Microwave radio transmission systems	23
2.7	Relationship between the transmission network and traffic or service networks	24
2.8	Switching and signalling developments leading to the integrated digital network	26
2.9	Mobile radio networks	31

Chapter 3 The transmission of data and other non-telephone traffic 35
    3.1 Data, facsimile and video traffic on the telephone 35
        network
    3.2 Telex networks and teletex 39
    3.3 Packet-switched data networks 42
    3.4 Data interfaces, protocols and open systems 46
        interconnection

Chapter 4 The integration of public networks and services 50
    4.1 Local-network developments leading to the 50
        integrated-services digital network
    4.2 Integrated speech and data: the service-independent 54
        network concept

Chapter 5 Private networks 58
    5.1 Leased private circuits on public networks 58
    5.2 Private branch exchanges (PBXs) and private 60
        telephone networks
    5.3 Private data networks and the local area network 65
        (LAN)
    5.4 Integrated speech and data on PBXs and LANs 75

✔ Chapter 6 Traffic measurement, forecasting and equipment 78
        quantities
    6.1 The purpose of traffic measurement 78
    6.2 Telephone-network traffic measurement and 79
        forecasting
    6.3 Message-switched data-network traffic measurement 84
        and forecasting
    6.4 Packet-switched data-network traffic measurement 85
        and forecasting
    6.5 From forecast to equipment quantity: the effect of 87
        grade of service, circuit occupancy and network
        delay

Chapter 7 Traffic considerations in switching-system choice and 96
        dimensioning
    7.1 What is a switching system? 96
    7.2 Telephone switching systems and PBXs 97
    7.3 Data switching systems 103
    7.4 Processor control 105
    7.5 Switching systems and traffic routing 109
    7.6 The switching of integrated speech and data traffic 113

Chapter 8   Traffic considerations in network planning   115
    8.1   Telephone networks   115
    8.2   Data networks (excluding local area networks)   125
    8.3   Local and wide area networks   129

✔ Chapter 9   Traffic-carrying performance evaluation   132
    9.1   The role of the performance engineer   132
    9.2   Performance evaluation methods   135

✔ Chapter 10   The effects of traffic-carrying performance and   139
    public-network tariffs on private-network costs
    10.1   Traffic-carrying performance and public-network   139
    tariffs in the context of private-network costs
    10.2   The design of private telephone networks to a   140
    performance target
    10.3   The design of private data networks to a   145
    performance target
    10.4   Public-network tariffs   147

Chapter 11   Finding the right system/network solution   154
    11.1   Establishing the requirement   154
    11.2   Narrowing the choice   155
    11.3   Finding the most cost-effective solution   158
    11.4   Cost-comparison examples   161

Bibliography   177
Appendix A: Full-availability traffic table for various grades of   179
service
Appendix B: High-usage leased circuits with overflow to the public   181
telephone network

Glossary   182

Acknowledgment

This book would not have been written but for the encouragement of Professor Charles Hughes of Essex University, formerly of the British Telecom Research Laboratories at Martlesham Heath, Ipswich. I would also like to record my debt to Colin Smith for his helpful comments, to many other former colleagues who contributed to my enjoyment of a career in telecommunications spanning more than 40 years and, above all, to my wife for her continuing support.

R.E. Farr

Introduction

Telecommunications is an all-embracing term, covering the interchange of information in the form of speech, pictures or data. Originally, the definition of telecommunications required that the interchange should occur by electrical means but, with the introduction of optical fibre as a transmission medium, the means can also be optical. Information is exchanged between people or computers – *telecommunications users* in the context of this book – over the circuits of a public network, or over circuits that may be leased from a public carrier or privately owned.

The information being exchanged, whatever its form or mode of transmission, will occupy the circuits over which it passes for a finite period in the way that moving traffic occupies a road. Thus, when information is being passed over a telecommunications circuit, the circuit also is said to be carrying *traffic*. All forms of telecommunications information flow are covered by the term *telecommunications traffic*. To understand the fundamental properties of telecommunications traffic and the differences between one type of traffic and another, it is necessary to know in outline what actually happens when, for example, one user makes a telephone call to another or data is exchanged between terminals.

Telecommunications users must be connected by a continuous transmission path for information to pass between them. When there are many users, it would clearly be uneconomic to provide every one of them with a separate path to every other user; hence a means must be provided for sharing the paths and switching between them as required. Any arrangement of telecommunications equipment that provides such information paths may be described as a *telecommunications network*, whether there are a small number of or several million users. A telecommunications network has a finite capacity for transferring information. The constituent parts of a network – telephones and data terminals etc., the private branch exchanges (PBXs) and public exchanges (central offices in the USA) which carry out the switching function, and the circuits that provide the transmission paths between them – may be considered as building blocks. Each building block has its own individual finite capacity for handling information, which should be tailored to match the amount of traffic it will be required to carry. Too much capacity will mean added expense; too little will cause traffic to experience delay or fail to materialise because a connection cannot be established. Both situations can exist in different parts of the same network.

The main function of PBXs and public exchanges is to switch between a number of circuits to different destinations – hence the all-embracing term *switching centre*. Whether the switching centre is large or small, it is built up from a number of equipment modules and circuits. These modules and circuits are the constituent building blocks of a switching centre which is itself a network building block. Some of these constituent building blocks are supplied to order, others are built in by the manufacturer in a fixed relationship with each other. Each must perform to a standard which, when all are assembled in the correct relationship, ensures that the required amount of traffic will be carried efficiently at an economic price.

The manager who, for example, wishes to invest in a new **PBX** can choose on the basis of *cost* alone between a large number of different products which all offer the required facilities. It is also important, however, to know whether these products will *perform* to a comparable and adequate standard. This requires two important questions to be asked of each product under consideration:

(a) Will it be capable of providing the rquired traffic-carrying capacity now and in the future?
(b) How many 'available to order' equipment modules and circuits are really necessary to match the needs of the business?

The same questions are equally applicable to, for example, a large data network and its constituent parts. Insufficient capacity will mean delay in transferring information, which may prove expensive for the future running of the business; too much capacity will mean wasted capital expenditure now.

Decision making today is made more difficult by the enormous diversity of telecommunications products available, and the fact that modern technology breeds products whose inner workings and performance capability are not apparent to the purchaser or user. A choice may need to be made between providing private facilities or making use of a public telecommunications service. All too often, where a business does not employ a professional engineer to manage its telecommunications facilities, vital decisions have to be made in ignorance of the basic principles involved and on the basis of external advice that may well be partisan.

The primary aim of this book is to equip managers who are called upon to make these expensive decisions on their telecommunications needs with sufficient information to be able to judge those needs and the possible solutions on an informed basis. To this end, the book explains the fundamental properties of telecommunications traffic, how telecommunications systems and networks transfer information, and why it is essential that they are designed and assembled with traffic-carrying performance in mind if they are to provide an efficient service at minimum cost. The basis of public-network tariffs and their effect upon network costs are also considered. Explanations and examples are kept to as simple a level as possible to encourage a wide readership. It must be accepted, however, that telecommunications is a very complex subject. No book that sets out to achieve the objective of informing managers about telecommunications in the compass of relatively few pages can do so without leaving out some aspects and simplifying others, to a degree which the technical purist might fault. The book is not, therefore, intended as a do-it-yourself guide for the uninitiated. No one book can be expected to provide a solution to all the complex networking problems likely to be encountered in practice, particularly in the field of data communications; ultimately, there is no substitute for the advice of a qualified and preferably independent consultant.

Chapter 1

The nature of telecommunications traffic

1.1 Telephone traffic

When the handset of a public-network telephone is lifted, the circuit between telephone and switching centre – known as the *local circuit* – is immediately barred to potential incoming callers and is said to be *engaged* or *busy*. Even though a connection to the intended destination has yet to be established and no information in the form of speech can be exchanged, to all intents and purposes the local circuit is already carrying traffic since it cannot be used for any other call. The initial purpose of the local circuit is to convey signals representing the dialled or keyed digits to the switching centre, so that they may be interpreted in order to extend the path for speech communication to the required destination. Only when a complete end-to-end circuit has been established between the calling and called telephones can the local circuit be used for its intended purpose of exchanging information in the form of speech – the originally introduced concept of what is meant by telecommunications traffic. The sounds of speech are represented by variations imposed on a steady electric current, and it is these current variations which are conveyed as a continuous signal between the telephones to provide two-way communication.

Two different concepts of *telecommunications traffic* have now been identified:

(a) The *overall* concept that a circuit is carrying traffic when it is engaged and unavailable for any other call, whether or not information is being exchanged.

(b) The more *limited* concept of a circuit carrying traffic when information – in this case, speech – is able to be exchanged between two telecommunications users.

From the point of view of assessing the capacity of a *switching centre* or *network* to handle a given number of calls, the important factor is whether there are free circuits available to take new calls; thus the overall concept of traffic (a) is appropriate. When the term 'traffic' is used without qualification, it is this overall concept that is implied. A telephone call actually comprises three elements of time: the initial connection or *set-up* period, the period during which information can be exchanged, and the final disconnection or *clear-down* period. The limited concept (b) excludes both set-up and clear-down periods; it covers only the period during which a conversation can take place. This element of traffic is, therefore, identified as *conversational traffic*.

The overall duration of a call from the beginning of set-up to the end of clear-down is known as the *holding time* of the call, and is one of the main parameters to be considered in the measurement of traffic. Holding time is a general term that may also be applied, for example, to the engaged time of a piece of equipment not in use for the full duration of a call; thus switching-centre equipment in use during call set-up only will have a very brief holding time which bears no relationship to the holding time of the call being set up. When holding time refers specifically to call duration, therefore, it is preferable to use the term *call holding time* to ensure that the meaning is clear.

So far, this discussion has been limited to the traffic, and thus the call holding time, that would be measured on the local circuit between the caller's telephone and its parent switching centre. Where the caller has access to the public network via a PBX, a connection is established across the PBX to a local circuit between the PBX and a public-network switching centre, and it is the traffic on this local circuit that is of importance from the public-network point of view. Although a local circuit may be shared in special circumstances, it is normally provided for the exclusive use of a single *customer* – a term which covers the complete range of users, from those having only one directly connected telephone to those grouped together to be served via a PBX. The local circuit can only be used for either an outgoing call from or an incoming call to that

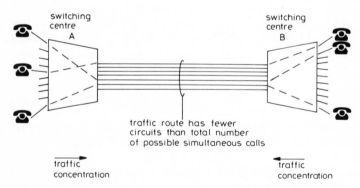

Fig. 1.1 *Sharing circuits*

customer at any one time. The switching centre which routes calls to the required destination also serves as a traffic concentration point, however, so that between switching centres there will always be circuits that are provided on a shared basis for the use of a number of customers (Fig. 1.1). A group of such circuits between two switching centres in a network is referred to as a *traffic route.* The call holding time of a particular call as measured on a traffic route will be slightly shorter than the holding time of the same call as measured on the caller's local circuit. This arises because the first part of the set-up period, from the time that the caller's handset is lifted until the dialled or keyed routing digits have been acted upon, is confined to the caller's local circuit and the switching-centre equipment associated with it. Thereafter, the residue of the overall set-up period becomes shorter as set-up progresses through the network towards the required destination (Fig. 1.2).

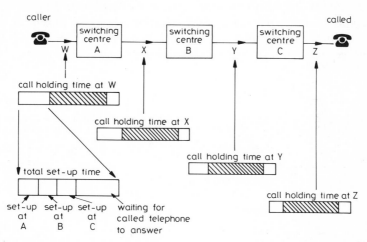

Fig. 1.2   *Breakdown of call set-up time*

It is important to understand that whereas all telephone circuits must be capable of exchanging information in both directions simultaneously, call set-up can only occur in one direction at a time. The circuits of a traffic route may be equipped to enable calls to be set up over them (a) in one fixed direction only, when they are referred to as *unidirectional* circuits, or (b) in either direction on a first-come, first-served basis, when they are known as *bothway* circuits. The circuits of a traffic route are selected automatically, in a predetermined order, as each new call needs to be set up over the route. One of the circuits will be seized for a call and, when the conversation has ended, the circuit will be cleared down and become free for another call. In this way a succession of calls from different users may pass over the circuit; where the circuit is bothway the set-ups will occur in either direction indiscriminately. This idea of a succession of calls passing over a circuit gives rise to the important concept of *traffic flow*.

1.2 The effect of unsuccessful calls on telephone traffic flow

Telephone traffic flow has so far been explained in terms of successfully completed calls but some call attempts, indicated to the caller by an engaged tone or a recorded announcement, are not successful. The distant telephone may be engaged, it may prove impossible to find a free circuit *en route*, or the circuit-selection process may fail because of an equipment fault. An unsuccessful call attempt will, nevertheless, engage some circuits as far as the point where the attempt fails; these circuits will not be available for other calls until clear-down takes place (Fig. 1.3). The holding time of an unsuccessful call attempt still consists of three periods, but these are now the set-up time to the point where the attempt fails, the waiting time until the caller decides to abandon the attempt, and the time to clear down the engaged circuits once the caller replaces the telephone handset. The set-up time will vary in length according to how far the call attempt has progressed before failure occurs and the overall holding time can be expected to be shorter than for a successful call.

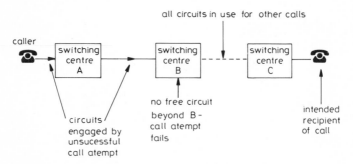

Fig. 1.3   *An unsuccessful call attempt*

Traffic flow can be quantified by adding the holding times of all calls using a circuit in a given period. If some call attempts offered to a traffic route fail to find a free circuit, they are said to be *blocked* and the route is said to be *congested*. The number of call attempts offered to the route will be greater than the number of calls successfully carried by the route. The call attempts that fail represent traffic that would otherwise have been carried by the route and are referred to as *lost traffic*. There is lost traffic throughout any telephone network because there will always be some call attempts that fail. If it is assumed that each call has the same average holding time, it then becomes possible to quantify flow in terms of the number of calls; thus *average traffic flow* equals the number of calls multiplied by the average call holding time. If it is now assumed that each call that fails would have lasted for the average call holding time had it been successful, it is possible to put a value on the lost traffic: this will be the number of failed call attempts times the average call holding time. Two other terms are introduced on the same basis: *offered traffic* is the total number of call attempts offered to, for example, a group of circuits, multiplied by the average call holding time; *carried traffic* is the number of calls actually carried by that group of circuits multiplied by the average holding time. Then

$$\text{traffic offered} - \text{traffic lost} = \text{traffic carried}$$

It should be noted that offered traffic and lost traffic are purely hypothetical quantities which are introduced to permit the mathematical analysis of traffic-carrying performance, whereas carried traffic is a real quantity.

1.3 Traffic-flow units and circuit occupancy

A unit of traffic flow is correctly defined as the total time a circuit is in use during a known measurement time, divided by that measurement time. Strictly, the total time a circuit is in use – the sum of call holding times – should be regarded as representing the *volume* of traffic using the circuit, expressed as a period. Traffic flow is, therefore, really a measure of traffic volume over time. The internationally recognised unit of traffic flow is based on the concept of measurement over 1 hour, and is named the *erlang* (unit symbol E) after a Danish pioneer in telecommunications traffic theory, A.K. Erlang (1878–1929). It is expressed as a decimal figure with a maximum of 1.0 per circuit. A circuit in use for three-quarters of the measured hour is, therefore, carrying 0.75 E of traffic. Measurements may be considered in minutes provided the measurement time is taken as 60 minutes (45/60 = 0.75) or seconds provided the measurement time is taken as 3600. The total traffic flow on a traffic route will be the sum of the individual traffic flows on each circuit; traffic flow can also be expressed in terms of an average per-circuit figure. It may be of interest to note that, since traffic flow is quantified as a measure of time divided by a measure of time, the erlang is an 'abstract' number which has no dimension in reality.

The sum of call holding times for a given circuit can also be considered in terms of the proportion of the measured hour the circuit is engaged or 'occupied' – so giving rise to the term *occupancy*. Occupancy is a measure of the efficiency of circuit usage, and is also expressed as a decimal quantity with a maximum value of 1.0. Thus, the occupancy of a circuit is numerically equal to the traffic flow on that circuit in erlangs. Occupancy is always considered on a per-circuit rather than a cumulative basis; thus the occupancy of a group of circuits will always be expressed as an average figure.

In the USA and some other countries which follow American practice, the erlang is not universally accepted as the unit of traffic-flow measurement. The alternative American approach to traffic measurement requires the time a circuit is in use to be measured in multiples of 100 seconds instead of as a decimal part of 1 hour. This gives rise to the title *hundred call seconds per hour* (CCS) for the American unit. The traffic flow on a continuously occupied circuit would be (3600 seconds)/100 = 36 CCS = 1 erlang.

1.4 Loss or delay working and repeat attempts

The method of handling traffic in which a call attempt fails or, in other words, is lost if a free circuit cannot be found immediately during set-up, is known as *loss working*. However, it is possible to design the circuit-seizure operation so that set-up will be automatically delayed until a free circuit is available; in effect, the call attempt is made to wait in a queue. This method of handling traffic, where set-up is delayed instead of a call attempt being lost altogether, is referred to as *delay working*. With delay working, there will be a fixed maximum queue size to limit the number of waiting call attempts. When the queue is full, subsequent call attempts will be lost – that is, the system reverts to loss working – and when a call attempt already in the queue succeeds, to leave a vacant queuing space, delay working will be resumed. Most telephone switching centres, in fact, incorporate elements of both loss and delay working in their design. Set-up delay should not be confused with the cross-network delay that results from data network store-and-forward working (Section 1.8).

A variation of delay working enables a terminal or switching centre automatically to make a *repeat attempt* to set up a call (sometimes referred to as a retry) if a free circuit is not found at the first attempt. Many modern telephones have a 'repeat last number' button which enables a caller to repeat the number last keyed should a call attempt fail. An automatic repeat-attempt feature for outgoing circuit selection is commonly incorporated in telephone switching-centre design; this operates without the caller having to take any action or even being aware that there has been a first-attempt failure. One automatic repeat attempt only is normally catered for, although this depends upon the speed of operation of the switching system in use. This restriction is imposed because several switching centres will be involved in setting up a long-distance call and the average caller expects to have to wait only a limited time for an indication of success, or failure, after dialling or keying the required number. Should the second attempt also fail the system reverts to loss working and the call attempt is lost.

The overall time needed to set up, or attempt to set up, a connection between any two users on a telephone network includes the dialling/keying and switching time plus any additional time resulting from delay or repeat-attempt working in the network. A limitation on the amount of delay working generally, and on the frequency of repeat attempts in particular, is important to the network operator because every queued call attempt or unsuccessful repeat attempt adds to the non-revenue-earning traffic flow on the network. In both cases, circuits preceding the point of queuing or call-attempt failure are being prevented from carrying calls that could otherwise be completed successfully. Since these set-up delays will mostly be incurred if the network is already busy, it is necessary to ensure that congestion is not further increased by a combination of delay-working queues that are full and too many unsuccessful repeat attempts in succession. This is especially true of repeat attempts originated at a terminal, since a large number of terminals may be generating these at busy periods. It may be argued that in most of the cases where the automatic repeat-attempt facility is invoked, a manual repeat attempt would have been made anyway, but manual attempts could never be as rapid or persistent and could not, therefore, overload a network in the same way.

1.5 Variability of traffic flow: the busy hour

A major problem in determining, for example, the appropriate number of circuits for a particular traffic route is the variable nature of telecommunications traffic. The demand for circuit use is originated on a random or 'pure chance' basis and traffic flow will, therefore, vary considerably from minute to minute and hour to hour. Nevertheless, some pattern of traffic flow throughout an average day will emerge in every telecommunications network; in telephone networks, for example, there will usually be a period of very low demand during the night and one or more periods during the day when demand is at its highest. Given a large enough population of telecommunications users, the timing of these periods and the volume of traffic flow will be remarkably consistent. This gave rise to the idea of the *busy hour* – the hour during an average day, calculated from traffic measurement, when traffic flow can be expected to be at its greatest. This busy-hour demand has to be met on a consistent basis and, therefore, determines how many circuits should be provided. Although there may well be other periods during the day when traffic flow is also at a high level, there will also be times when traffic flow is at a very low level indeed and the circuits are considerably under utilised.

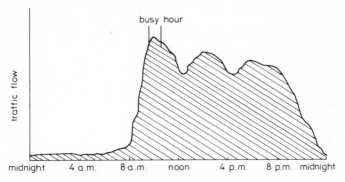

Fig. 1.4   *Typical daily traffic flow on a public telephone network*

The network busy hour in a large public telephone network will usually occur when business activity is at its peak during the morning of a working day (Fig. 1.4). Tariffs play a significant part in levelling demand during the day; thus, reduced afternoon rates will be used to encourage a spread of business demand and cheap evening rates to promote social calls at a time when business activity is minimal. Each switching centre in the network will have its own particular busy hour, varying from that of the network by a different amount, depending on the particular mix of business and residential needs served by that centre. In addition, outgoing traffic from a switching centre may predominate at certain times of day and incoming traffic at other times, giving rise to different busy hours for the various modules within the switching centre – the building blocks referred to in the introduction to this book. Each traffic route connected to a switching centre will also have its own particular busy hour which may differ from those of other routes and the switching centre itself. This is particularly noticeable with international traffic where time zone differences have an important effect on the timing of peak traffic periods. The demands imposed by a large number of *non-coincident* busy hours at each switching centre have to be met by the provision of equipment and circuits in the most efficient and economic way possible.

Although the amount of traffic flowing during the busy hour is likely to be reasonably consistent from day to day (excluding weekends when the pattern changes and the volume of traffic can be expected to be less), some switching centres and traffic routes may experience a peak demand on a particular day each week. Another common occurrence is the seasonal peak demand found where switching centres or traffic routes serve holiday areas. Overriding all of these daily, weekly and seasonal variations will be a general trend in growth or decline of traffic, determined by the economic health of the area being served.

All these variations are to some extent predictable in terms of the average level of traffic flow, but it is inevitable that there will be random fluctuations from this average. Occasionally, unforeseeable circumstances will arise – for example, a major accident or natural disaster – which can raise the level of demand for telecommunications services dramatically, well above the forecast average and therefore well above the network capacity that has been provided.

Similar patterns of changing demand will also apply to private telecommunications networks, the timing of the busy hour depending on the nature of the business. The smaller number of users tends to reduce the accuracy of forecasting, however, and random fluctuations in traffic flow are likely to have a more significant effect than on public networks.

1.6 Grade of service

The forecast average level of traffic flow at the busy hour is commonly used as the basis for determining (a) the number of circuits to be provided on a traffic route and (b) the number and size of switching-centre modules. Predictions will be derived from measurements taken over a large number of busy hours for the particular route or module concerned. Taking the example of a telephone-network traffic route, if sufficient circuits are provided to give a traffic-flow capacity which exactly matches the forecast busy-hour traffic flow, such that theoretically each circuit will be fully occupied throughout the busy hour, an unacceptably high proportion of call attempts will fail. This happens because calls are originated at random times and do not arrive in a continuous, uninterrupted stream; in practice, a considerably greater, *uneconomic* provision of circuits would be needed to ensure 100% success. A means has to be found of relating a realistic level of circuit provision to the predicted traffic flow, in order to provide an adequate service at an acceptable price.

The concept of *grade of service* (GOS) was introduced to provide this relationship and is particularly applicable to telephone networks. GOS is, as the name implies, a measure of the quality of service a user may expect when making a call during the busy hour, but it is specifically limited to quality in terms of the sufficiency of circuits and switching-centre equipment provided in the network. No account is taken of the possibility that circuits or equipment might not be available because of a fault. GOS is specified in terms of the probability that a given proportion of call attempts offered, for example, to a traffic route, will be lost; it is usually expressed as a decimal figure. Thus, if 1 call attempt in every 100 can be expected to fail owing to a shortage of circuits or switching-system equipment, the GOS is expressed as 0.01; a GOS of 0.02 means that 1 in every 50 can be expected to fail, and so on. It is then possible to specify a particular GOS and calculate the number of circuits required to meet this specification. The determination of an appropriate GOS for the provision of circuits or equipment in a particular situation involves the weighing of user satisfaction against economic factors, as discussed in later chapters.

1.7 Data traffic

It is appropriate at this stage to restate the description given of a telephone call in a more general telecommunications sense: the call is a complete, self-contained sequence of events which enables information to be exchanged between two telecommunications users. The sequence embraces not only the conversational traffic period but the set-up and clear-down periods as well, since the conversation cannot take place without these. In other words, a call is defined as including the information being exchanged between two telecommunications users and also the information needed to bring those users into contact and separate them afterwards. This more general understanding can then be applied to other types of telecommunications traffic.

Data communications may be considered as the exchanging of information between machines rather than people. Although early data systems required the direct intervention of people, modern user terminals may have computer intelligence and the ability to store information which render this unnecessary. It is important, however, to note two fundamental differences between telephone and data traffic. First, for a successful telephone call, it is essential that the exchange of information should be complete and uninterrupted; unless this requirement is met, an intelligible conversation cannot take place. Secondly, a telephone circuit must be capable of passing information in both directions simultaneously. With data traffic, intelligibility need not be impaired if there is a break or delay in the exchange of information because the data is transferred between machines in a precisely controlled way, and it is not essential to exchange information in both directions at the same time.

Data traffic originated with telegraphy, which predated the telephone. Early telegraph circuits were capable of transmitting messages in one direction only and became known as *simplex*. Later, *duplex* circuits were developed to exchange messages in either direction simultaneously, and the term *half-duplex* was applied to circuits capable of transmitting either way but in one direction only at a time; these terms are still widely used to describe data circuits. The essential feature of telegraphy was the use by operators of a code to represent letters and numbers by different combinations of pulses of current, and coded pulses are still the basis for all forms of data transmission today.

The original public telegraph service required the delivery of a message from one user to another by a third party; this was followed by the telex service, which introduced the concept of direct user-to-user communication. The calling telex operator would input a message using a typewriter-style keyboard, and coding into pulses for transmission over a line was carried out by an electromechanical teleprinter or teletype. The receiving machine interpreted these coded pulses as characters, printing them as typescript. Today, many telex services use an independent network operating on the store-and-forward basis described in the next section, and modern telex terminals incorporate electronic circuitry and computer control.

In effect, terminals are able to 'talk' to each other over a telex network in much the same way that people would. Telex traffic is made up of separate interchanges which are effectively complete 'calls'. Data traffic sent over the public telephone network is also in the form of complete calls, the set-up and clear-down procedures being those used for telephone calls. In both cases, therefore, traffic flow can be measured in erlangs or CCS as for telephone traffic.

1.8 The store-and-forward concept and delay

A delay in the receipt of data information need not affect its intelligibility, as would be the case with a telephone conversation. This feature of data traffic can be used with advantage since random peaks of traffic can be smoothed out by delaying some messages when traffic is at its heaviest (Fig. 1.5). How long a

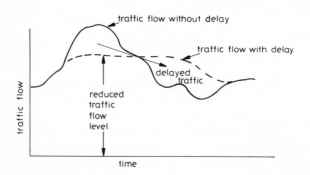

Fig. 1.5 *Delaying data traffic to smooth traffic flow*

delay can be tolerated depends very much on the user's requirements. Where, for example, a computer file is being interrogated for an urgently needed item of information, seconds are important, whereas several hours delay may be unimportant in other cases. Delay can be introduced within the network itself or at the data terminal.

Data messages sent over a telephone network cannot be delayed by the network in this way, because it is randomly shared by telephone conversations which must not be interrupted by delays. With a purpose-built network, however, data messages can be automatically stored at a data switching centre until an appropriate free circuit is available (Fig. 1.6). *Store-and-forward* working

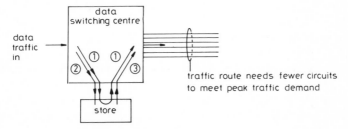

Fig. 1.6   *The store-and-forward concept:*
1 Where free circuit is available, traffic is stored and retransmitted immediately
2 Where no free circuit is available, traffic is held in store
3 When free circuit becomes available, traffic is retransmitted from store

requires data messages to carry their own identity and destination information so that the switching centre can direct them in sequence to the required destination. The delay introduced by a switching centre is a function of the network, outside the control of users, and affects the transit time of a data message across that network. It can be shown that the shorter the average holding time, the shorter will be the average network delay; hence messages in each direction are treated as separate calls. It is, in fact, the measurement of this network delay, to ensure it is within limits acceptable to the users, that assumes the greatest importance in assessing the traffic-carrying performance of these networks. Practical examples of the store-and-forward principle – permitting network circuit capacity to be used with greater efficiency – can be found in modern telex networks, where message length can vary widely and network delay is not particularly critical. It is still possible to measure traffic flow in erlangs or CCS, but holding times are on an individual message instead of a call basis.

The fact that store-and-forward data networks are designed to accept data immediately, without the assurance of a through circuit to the recipient, obviates the need for the repeat-attempt circuit selection procedures common in a telephone network. At times of heavy traffic congestion, however, delays to the transfer of data could become excessive if no limitation were imposed. In practice, the restricted storage capacity available will impose such a limitation;

when the storage at a switching centre is full, no more traffic can be accepted until some data in the store can be successfully transmitted, to leave a vacant space.

Data terminals are also obtainable with a built-in store-and-forward capability. A non-urgent message can be input during business hours when the network is heavily loaded and, for example, automatically transmitted later when demand is slack (on public networks, this also permits advantage to be taken of cheaper off-peak rates). Here, the user at the sending terminal has control over the introduction of delay and, since transit time across the network is unaffected, terminals designed for use on either a data or a telephone network can be so equipped.

1.9 Packet-type data traffic

The latest public data networks, designed for the rapid response and high-speed data transfer necessary where a computer is being interrogated, utilise a development of the store-and-forward concept which employs a different principle for transmitting information. This principle, also extensively used in private networks, is explained more fully in Section 3.3. In essence each data message is cut at very short, fixed intervals into a number of sequential parts known as packets (Fig. 1.7). These message-carrying packets are transmitted

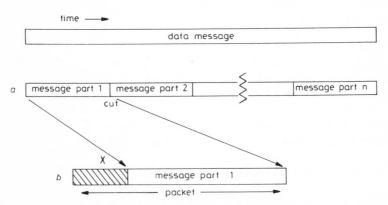

Fig. 1.7   *The data packet concept:*
(a) Data message cut into parts at fixed intervals
(b) Packet formed from each message part by adding identity, sequence and destination information X. Each packet can then be transmitted to destination independently

independently, carrying their own identity, sequence and destination information in addition to part of the message. Each packet, therefore, has a self-steering property and can be handled as if it is a message in its own right. Packets are interposed with the packets of other messages and do not necessarily have to use the same fixed routing to reach a given destination; these features allow circuits to be used very much more efficiently than would be possible with a conventional telephone network. The very short packet holding time ensures that network delays are minimal. The packets are ultimately stored and reassembled in the correct order to reproduce the complete message to the intended recipient.

The packetised data concept requires the idea of call holding time as applied to a telephone call to be reconsidered; each packet is, in effect, a call in its own right but carries only part of a message, which is itself only part of a call in the telephone-network sense. The holding time of each packet is both very short and fixed; there is no point, therefore, in considering traffic flow in erlangs or CCS when a count of the number of packets transferred in a given time provides an adequate means of quantification. The most important factor from the user's point of view is the speed of user-to-user data transfer; this is governed by the incidence and length of network-induced delays.

Chapter 2

The transmission of telephone traffic

2.1 Signalling and analogue transmission over the local network

Public telephone network local circuits, as introduced in Chapter 1, together with the cables etc. which carry them, constitute the *local network*, and the switching centre to which they are connected is referred to as a *local switching centre*. Each local circuit will usually serve one customer and must normally consist of a pair of electrically conducting wires for speech transmission purposes. These same two wires are used to signal the calling condition and convey the dialled or keyed digits to the local switching centre, functions to which the generic term *signalling* applies (where a telephone is an extension on a PBX, additional wires may be provided for this purpose). The explanation of the transmission of telephone traffic given in the following paragraphs is particular to a metallic-pair local network, but radio or optical-fibre transmission systems may be appropriate in particular circumstances and can be expected to become an increasingly acceptable alternative in the future (see Sections 2.9 and 4.1).

Speech causes the surrounding air to vibrate and it is these air vibrations – so-called sound waves – which are detected by the human ear. The speed or *frequency* of vibration in *cycles per second* (c/s) determines the pitch of the sound. The telephone microphone converts these air vibrations directly into a varying electrical current, which varies at the same frequencies as the air vibrates. This is known as *analogue transmission* because the current variations are an analogy of the sound waves. The unit of electrical frequency, also measured in cycles per second, is the *hertz* (unit symbol Hz), where 1 Hz equates to 1 c/s of sound. Although the human voice produces sounds over a wide range of frequencies, it is possible to recognise and understand speech adequately by selecting only those frequencies within a range of about 300 to 3400 c/s. Telephone instruments and, more particularly, public telephone network circuits are not, therefore, normally designed for the transmission of frequencies outside this range of about 3 kHz.

The current variations which represent speech can be transmitted over a lengthy electrical circuit at a speed which is, to all intents and purposes, instantaneous. However, the current variations are affected by distance, suffering both a reduction in amplitude (equivalent to sound volume) and increasing distortion, collectively known as *transmission loss*. The resulting impairment of sound waves and quality eventually leads to an inability to understand what is being said.

The source of the electric current on which the variations are initially conveyed is the local switching centre. All PBXs and some special facility telephones, however, require mains power as well, and cordless telephone handsets are linked to the telephone base by a low-powered radio channel, using rechargeable batteries as their power source. This same current is also commonly used to signal the dialled or keyed digits to the switching centre as sequences of current pulses; for example, two pulses represent the digit 2 and ten pulses the digit 0. This method of signalling a number, referred to as *direct-current signalling*, was designed to match the speed of operation of a dial to the speed of operation of electromechanical switching systems. However, the fact that it takes longer to transmit ten consecutive pulses than two results in an unnecessary delay where faster-input keyphones are working to a modern high-speed electronic switching centre. An alternative faster signalling method, known as *multifrequency (MF) signalling*, permits advantage to be taken of these higher operating speeds; this, however, requires a specially equipped keyphone and access to a special receiver at the switching centre. A digit is represented by two different frequencies, which can be heard as tones; these are transmitted simultaneously, each digit from 1 to 0 using a different combination of frequencies. Simultaneous instead of consecutive transmission of the signals for each digit permits a number to be received at the switching centre as fast as it can be keyed in, thus considerably reducing the set-up time. MF signalling is commonly provided as standard for telephones on modern PBXs, and is available as an option in public networks where the local switching centre is suitably equipped.

2.2 Beyond the local network: long-distance analogue transmission

A call beyond the local network will involve a chain of at least two switching centres and their interconnecting *traffic routes* (Fig. 2.1). The last switching centre in the chain – that on which the required distant-end telephone is parented – must also be a local switching centre. Where there are intermediate switching centres in the chain, these will usually be dedicated trunk or toll centres, located to concentrate traffic on to large main traffic routes. Network design considerations, which embrace the location and purpose of switching centres, are discussed in Chapter 8.

Any long-distance telephone call will involve several, possibly very long, traffic routes in tandem. How is the transmission of speech over such long distances possible? Historically, a number of developments have contributed to a solution. The earliest was amplification, but its introduction brought added complication to the network (Fig. 2.2). First, amplification is normally possible in only one direction at a time so that a four-wire amplification circuit is necessary – a pair of wires for each direction. Secondly, amplification is only effective over a limited distance; this required amplifiers to be installed at intervals on long routes, with conversion between two-wire and four-wire

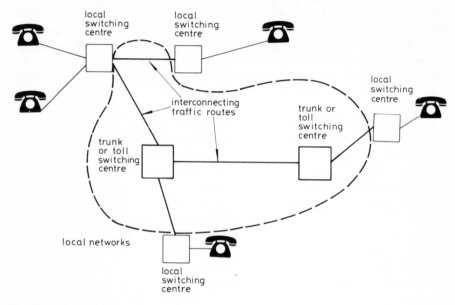

Fig. 2.1 *Beyond the local network.*

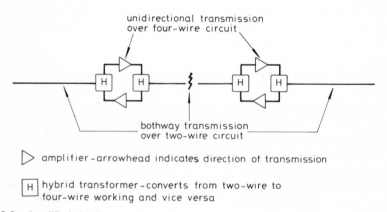

Fig. 2.2 *Amplified circuit.*

working at each amplification point. For the shorter routes – between two local switching centres, for example – an improved solution was later found with the invention of an inexpensive two-wire amplifier capable of amplification in both directions and effective over distances up to about 20 kilometres. This could be fitted at one end of a circuit, avoiding the need for intermediate amplifiers, and is still in common use. Although there is no direct electrical connection through an amplifier, bypass arrangements ensure that direct-current signalling is still feasible.

It is important to understand that, at best, amplification as used in telephony can only make up transmission loss; it cannot be used to increase the speech power above the original sending level because the circuit would then become unstable and inoperable. Since local circuits are not normally amplified, and the process of two-wire to four-wire conversion itself introduces transmission loss, there will always be a certain amount of transmission loss overall for any telephone connection; amplification merely limits this loss to an acceptable amount.

Amplification alone did not provide a lasting solution to the demand for ever-larger main traffic routes with the rapid growth of telephone networks. The next step was the discovery that cables are capable of carrying much higher frequencies. It is possible, in effect, to electrically lift the 3 kHz band of audio frequencies to an entirely different frequency range by imposing it on a higher *carrier* frequency. This process, also used in radio transmission, is called *modulation*. By using different carrier frequencies in a technique known as *frequency-division multiplexing* (FDM), a number of lifted audio-frequency bands can be accomodated on the same transmission path. Amplification still has to be provided at intervals, but the use of two separate transmission paths, one for each direction of transmission, enables the same range of higher frequencies to be used for each direction and requires conversion from two-wire to four-wire working at each end only. The unidirectional transmission paths, usually provided over a pair in each of two separate cables to avoid mutual interference, are termed *channels*. A complete *transmission system*, including the terminal equipment and intermediate amplifiers, provides a package for a fixed number of circuits; system capacity is stated in terms of the number of channels provided in one direction of transmission, and hence the number of channels quoted equals the circuit capacity.

Modern high-capacity transmission systems of this type are known as *FDM systems*. Two developments made these systems possible; (a) the introduction of *coaxial cable* able to carry very much higher frequencies with acceptable loss than pair-type cable, and (b) the development of high-speed electronic circuitry. Capacity is built up from a basic 12-channel group configuration by successive stages of multiplexing; the first stage is to combine five groups into a 60-channel supergroup, and thereafter European and US practice differs. FDM systems can provide from around 1000 to 10 000 circuits, operating at transmission speeds up to 60 MHz.

The gain in efficiency of cable utilisation made possible by these transmission systems is not achieved without cost and added complication, however. The higher the frequency transmitted, the greater the transmission loss over a given length of cable. This means that the interval between amplifiers has to be reduced compared with that for the transmission of audio frequencies; the largest systems require amplifiers every 1.5 km. In addition, the channels no longer provide the direct electrical connection previously regarded as essential for signalling purposes. The fact that signalling only takes place outside the conversation period, however, allows digits to be transmitted as pulses of an audio frequency – a method known as *voice-frequency (VF) signalling*. More recently, long-distance and international routes have used the more complex but faster multifrequency (MF) technique, similar to that described for local circuits in Section 2.1.

2.3 The need for international standards

It is appropriate at this point to consider communication between users on different national or intra-national networks. The worldwide public telecommunications network is the largest interactive machine developed by the human race and imposes a requirement for interworking standards on all concerned. The most obvious requirement is for compatibility of transmission systems and the signalling methods used with them. Where a system interconnects two separate networks, whether within or across national boundaries, the equipment must be compatible throughout the length of the route and must be able to interwork with other systems in either network. A second requirement concerns transmission performance; it must be possible to call any telephone in the world from any other telephone with the confidence that conversation will be clearly audible. This requires a maximum permissible transmission loss to be specified, not only for international circuits but overall for the worst-case connection within a national public network (including connections to and from private networks). Each network operator has to ensure that this maximum permissible loss cannot be exceeded, by imposing its own apportioned limits to the various classes of circuit in the network.

Although other standards authorities are concerned with particular areas, notably data communications and radio-frequency registration, overall responsibility for telecommunications standards rests with the International Telecommunications Union (ITU), a specialist agency of the United Nations based in Geneva. Responsibility is divided between two main committees, the International Telegraph and Telephone Consultative Committee (CCITT) and the International Radio Consultative Committee (CCIR), the standards being published as non-mandatory recommendations. Discussion is organised on a subcommittee basis, members being drawn worldwide from both network operators and manufacturers. A country will sometimes establish its own localised temporary standards, in advance of international agreement, to encourage the early introduction of new ideas into manufacture and field use.

2.4 Beyond the local network: digital transmission

The idea of transmitting speech in the form of coded pulses of current instead of as an analogy of the speech waveform first originated in the late 1930s. It was only with the development of high-speed electronic circuitry some 25 years later, however, that a practical system became possible. The process now adopted for such systems, known as *pulse-code modulation* (PCM), has a sufficiently wide-spread application to make an understanding of the principle involved desirable. The audio-frequency speech-signal waveform as transmitted over a local circuit is assumed, for the purpose of this explanation, to consist of a single tone at one frequency. This waveform is sampled for extremely short intervals at a very high sampling rate (Fig. 2.3). The pulses of current that result from this sampling process will have different amplitudes, the amplitude of consecutive pulses following that of the original speech waveform. Each sample pulse is in turn coverted into a number of even shorter pulses of uniform amplitude, the amplitude of the original pulse being indicated by the number and positioning of the shorter uniform pulses according to a predetermined code. These coded groups of pulses can then be transmitted to line, to be decoded and reconstituted as a replica of the original waveform at the distant end. Provided the sampling rate is fast enough (twice the highest audio frequency or around 8000 times per second), the resulting sounds are indistinguishable from the original speech.

By speeding up the pulse transmission rate prior to transmission, so that the individual pulses are of shorter duration, it is possible to interpose the pulses from a number of circuits – a principle known as *time-division multiplexing* (TDM). The pulses need to be detected and regenerated at intervals over long distances, just as analogue signals need to be amplified, the intervals decreasing with increasing pulse transmission rate. Since these signal *repeaters* are unidirectional in operation, the pulses for each direction of transmission must be channelled separately as in an FDM system. The pulses transmitted over each channel in the same direction of transmission, therefore, are cut at fixed intervals, each successive group of pulses being allocated the same fixed *time slot* in a repeated *frame* sequence; groups of pulses for the other channels occupy the remaining slots (Fig. 2.4). The pulses for all these channels are transmitted over a single cable pair, and a similar technique is employed for the channels in the reverse direction of transmission using another cable pair. The complete transmission system is known as a *PCM system*. Standard public-network PCM systems provide 24 circuits (the original UK and current 24-channel US standard) or 30 circuits (the current UK and European 30-channel standard) over two conventional cable pairs, normally over distances of between 10 and about 50 km.

Synchronisation of the pulse rates throughout a PCM system is critical to its operation and, although it would be technically feasible to transmit conventional VF signalling information over the system, it is more efficient to transmit both

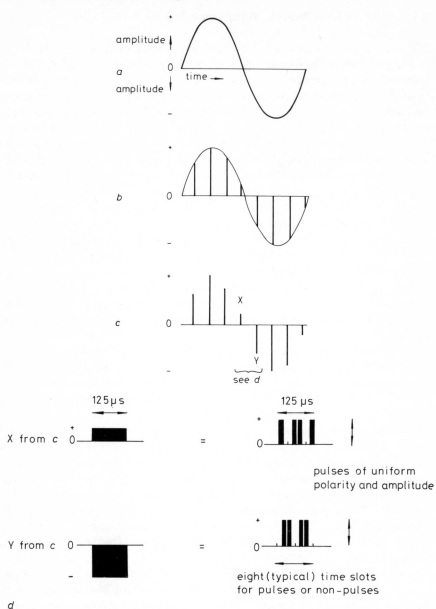

Fig. 2.3 *Pulse code modulation:*
Sampling (a) Speech waveform. (b) Sampling 8000 times per second. (c) Pulses represent speech waveform
Coding In (d), each pulse from (c) is replaced by a group of typically seven or eight uniform pulses or non-pulses within the same period (125 μs). One of the time slots takes account of the polarity of the pulse being coded (+ or − is pulse or non-pulse) and the remainder represent the amplitude, measured as being within one of a number of possible bands of amplitude level
Code conversion Time slots from (d) are converted to alternate + and − polarity to eliminate consecutive pulses of the same polarity before transmission to line.

synchronisation and signalling information directly as coded pulses. This is done within each channel frame in the 24-channel system, and in the 30-channel system within two spare channel time slots provided for this purpose (for this

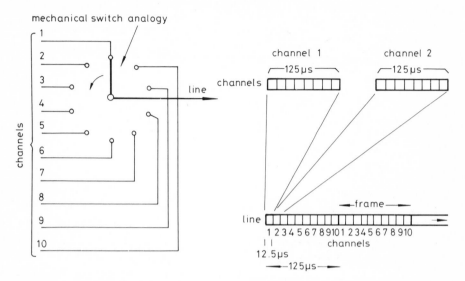

Fig. 2.4 *Time-division multiplexing. A mechanical switch analogy is shown: multiplexing is actually achieved electronically. Shown typically for 10 channels: the time allocated per channel is reduced by a factor of 10, i.e. sampling rate is increased 10 times.*

reason, the system is sometimes incorrectly referred to as a 32-channel system) (Fig. 2.5). Because PCM systems transmit information in pulse form, frequency – in the analogue system sense – is not involved. Speed of operation is, therefore, measured in terms of the pulse transmission rate, a pulse being known as a *bit* as in data transmission (see Section 3.1.1). A US 24-channel system operates at around 1.5 Mbit/s and a European 30-channel system at around 2.0 Mbit/s, each channel (the equivalent of 3 kHz analogue) being allocated 64 kbit/s.

Transmission over PCM systems is known as *digital transmission* because all information is transmitted in the form of numbers coded as pulses of current (compare this with data traffic as described in Section 1.7). It could be asked why the transmission of speech merits such a complex process when it has been transmitted adequately by analogue means for many years. The initial impetus for using PCM systems was twofold; other advantages will become apparent later in this chapter. First, PCM systems provide freedom from extraneous line noise and distortion and thus better transmission quality than analogue systems. Provided the pulses representing speech are capable of detection, they can be regenerated as new and the original speech can be exactly reconstituted. Secondly, PCM systems offer an economic solution to the demand for circuit growth on existing pair-type cables. A PCM system can be provided over two pairs which each previously carried only a two-wire amplified circuit, at less cost than would be the case with an FDM system of equivalent capacity.

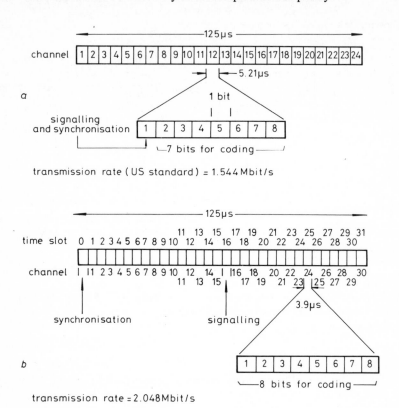

Fig. 2.5 *PCM frame structure: (a) 24 channel (b) 30 channel.*

Higher-capacity, longer-distance systems for use over coaxial cable have been developed from the original PCM concept by the use of multiplexing techniques, as with FDM systems. The digital systems, however, use successive stages of time-division multiplexing to speed up the transmission or bit rate and are known as *TDM systems*. Instead of achieving greater capacity by using an increased frequency range, time-division multiplexing produces the same effect by shortening the PCM time intervals so that more pulses or bits can be accommodated in the same timescale. Different ranges of systems have been developed from the basic US 24-channel and European 30-channel standards, operating at transmission speeds of up to around 500 Mbit/s to provide as many as 8000 circuits, but further development is unlikely with the advent of optical transmission systems (Section 2.5).

2.5 Digital transmission over optical fibres

The latest development in line transmission practice is the use of optical fibres instead of metallic conductors. Optical fibres make use of 'light' waves (usually of lower frequency than visible light) instead of electricity as the transmission medium. A typical fibre has a central core of silica glass, only a few thousandths of a millimetre in diameter, surrounded by a thicker concentric cladding layer of similar but more 'reflective' material (technically, the cladding has a higher refractive index than the core), the whole being not much thicker than a human hair. A high-intensity light source such as a laser is used to focus light into the end of the fibre core; the light waves are then transmitted along the core, being almost entirely contained within it by the more reflective cladding material and travelling for a considerable distance with minimal loss of intensity. Fibres are each encased in a protective plastic sleeve and made up into cables in a similar manner to metallic conductors.

PCM and TDM principles, as described in Section 2.4, are used to produce a succession of light pulses which represent the speech, signalling and synchronisation information for a group of circuits. On long routes the pulses have to be detected and regenerated at intervals, in much the same way as the electrical pulses transmitted over TDM systems. At the distant end the pulses of light have simply to be detected and used to reconstitute an electrical speech waveform. Optical transmission systems are based on the same 24-channel and 30-channel standards as TDM transmission systems.

The transmission loss along an optical-fibre cable is not as great as that along a coaxial cable. This reduces system costs by allowing optical repeaters to be spaced more widely along a cable than electrical repeaters; typically, for a 565 Mbit/s system, repeater spacing is increased from about 1 km to about 30 km. An optical system is also immune to all forms of electromagnetic interference. However, in view of the expected growth in demand for telecommunications services, the most important consideration is that light waves offer an enormous frequency range for future exploitation – a thousand times the whole radio-frequency spectrum – whereas TDM coaxial-cable systems are already at the limits of practicability. The cost advantage of optical systems accrues largely from the reduced need for repeaters; their future is, therefore, already assured for long routes where this advantage is most evident. For shorter routes, where the overall cost of optical systems is dominated by the currently expensive transmitting and receiving devices, PCM systems operating over metallic-pair cable remain the optimum choice. However, the use of cheaper devices – for example, light-emitting diodes in place of lasers – is already extending the range of possible optical applications.

2.6 Microwave radio transmission systems

Microwave radio is so called because of the very short wavelengths employed. Very short wavelength means very high frequencies, where radio transmission is by 'line of sight'; the transmitted radio waves are concentrated into a beam by means of a highly directional aerial aimed towards a similar receiving aerial which must be within unobstructed visual range. For this reason the aerials, normally of the familiar parabolic reflector or 'dish' type, are usually mounted on towers or convenient tall buildings. Transmission by means of microwave radio over long distances requires a chain of relay stations about 40 km apart, at each of which the radio signals are received and retransmitted.

Radio waves can be modulated in the same way as other electromagnetic waves; hence the principles that are applied to coaxial-cable line transmission systems also apply to microwave radio systems. These systems provide the equivalent of four-wire circuits, transmitting in each direction over a different frequency range; analogue systems use the FDM principle and more recent digital systems the TDM principle. The circuits provided over such a system are indistinguishable from those provided over cable routes, with signalling being carried out on the same basis. Microwave radio systems range in size from those designed to provide small numbers of circuits for island and other isolated communities to those capable of carrying large numbers of circuits on main traffic routes; typical of the latter is a digital system operating at 11 GHz (a gigahertz is 1000 MHz) which provides over 11 000 telephone circuits. Radio systems are particularly valuable where the laying of cable would be impracticable or too costly (this is also particularly relevant to private telecommunications networks), and as a supplement to established cable routes to give a network increased route diversity and hence security.

2.7 Relationship between the transmission network and traffic or service networks

So far in this chapter, discussion has concentrated upon a public telephone network as essentially a collection of switched circuits for speech transmission – either 3 kHz analogue or 64 kbit/s digital. However, traffic other than speech is also carried over these circuits: data, facsimile (the transmission of still images) and, to a limited extent, video (the transmission of moving images) (see Section 3.1). The transmission techniques and set-up and clear-down procedures are inherent to the network, so that all traffic carried must be compatible with these.

A public telephone network is actually made up of switching centres and the traffic routes that interconnect them. The local network, although predominantly used for telephone customers, is in reality a shared resource for all telecommunications customer connections. The microwave radio and line transmission systems which carry the traffic routes, together with the associated cables, are similarly regarded as a separate entity – the *transmission network* – because they are common to all telecommunications services and used to carry all types of traffic. A transmission network can be considered as providing *bearer* circuits

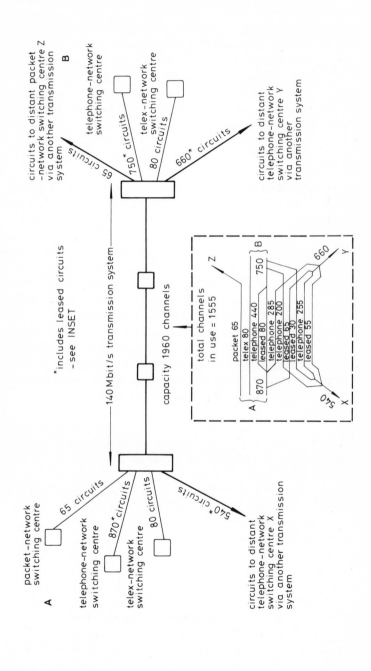

Fig. 2.6 *Aggregation of circuit demand on a transmission system.*

needed by the various traffic or service networks; these are circuits that provide 3 kHz analogue or 64 kbit/s digital paths and, where required, multiples or submultiples of these as explained in succeeding sections. Although large transmission systems have the capacity for many hundreds or even thousands of circuits, these do not all have to be equipped initially; circuit provision is modular, with capacity being increased by the addition of groups of circuits as required. Each traffic or service network requires a particular number of circuits for a traffic route between two of its switching centres; the needs of all networks over that section of the transmission network, together with leased private circuit requirements, are aggregated to determine the total transmission capacity requirement (Fig. 2.6).

2.8 Switching and signalling developments leading to the integrated digital network

The first requirement of a telephone switching centre is to switch a call attempt through to a traffic route which will carry that attempt to the required destination telephone. The conventional method of switching, whether it is electromechanically or electronically controlled, requires a metallic path to be physically switched between a number of alternative destinations. Because a source is switched to one of a number of spatially divided destination points, this is known as *space-division switching*. It is normally cheaper to switch circuits on a two-wire basis. Signalling information is transmitted over the circuit or channel being controlled, hence the term *channel-associated signalling*. With two-wire space switching, each link in the chain of transmission systems between switching centres is treated separately, the required customer's number being repeated over each successive link, possibly using different methods of signalling. This is known as *link-by-link signalling* (Fig. 2.7a). An alternative approach made possible by four-wire instead of two-wire switching is *end-to-end signalling*, in which the dialled or keyed digits are transmitted in stages, using the same method of signalling throughout; the customer's number then has to be transmitted once only (Fig. 2.7b). With this approach, used particularly on long-distance international routings, the reduced number of two-wire to four-wire conversions involved reduces transmission loss and, since there is no longer a need to repeat signalling information, set-up times are shorter.

Digital transmission systems by themselves do not alter this situation; the impetus for change comes from the introduction of high-speed electronic circuitry and computer control to switching systems. This has led to the concept of a network where speech is encoded for digital four-wire operation at the point of entry and remains in this form throughout its passage of the network, until it is decoded at the distant end. This requires the speech information to remain in digital form through each switching centre, and makes use of a different method of switching known as *time-division switching*. Local circuits and any other analogue circuits terminating on the switching centre are converted to digital working and multiplexed by PCM terminal equipment, appearing on the time switch as channels in 1.5 Mbit/s or 2 Mbit/s digital streams; digital circuits carried on PCM/TDM routes appear similarly. The information appropriate to a given circuit is then being carried in a particular PCM time slot. To connect this circuit to another circuit in a different PCM system, the information being

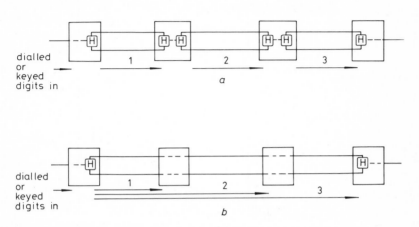

dialled
or
keyed
digits in

a

dialled
or
keyed
digits in

b

Fig. 2.7   *Channel-associated signalling. Signalling information (routing digits plus required customer's number) is transmitted over the speech circuit being set up*
(a) Link-by-link signalling. Signalling information is repeated over each link separately. Each link may employ a different signalling system suited to the type of circuit, but the staged progression of signalling results in lengthy call set-up times
(b) End-to-end signalling. Signalling information is transmitted in stages from initial switching-centre control equipment only. Since there is no repetition, call set-up time is reduced.

transmitted over the circuits must be shifted in time from one time slot to another (Fig. 2.8). The time shift can only be achieved by introducing a very small delay, unnoticeable in practice. Digital switching systems commonly incorporate stages of both space-division and time-division switching.

A network in which switching and transmission is integrated digitally in this manner is termed an *integrated digital network* (IDN). Such a network is free from the impulsive noise and distortion encountered in space-switched analogue networks, with significantly reduced end-to-end transmission loss and set-up times. Typically, digital encoding takes place at the entry to the local switching centre, and calls are then switched and transmitted digitally throughout the network until digital decoding occurs at the destination local switching centre (Fig. 2.9). All signalling takes place directly between the switching-centre control computers and, by dissociating the signalling information from the circuits to which it refers and concentrating it all on separate dedicated signalling

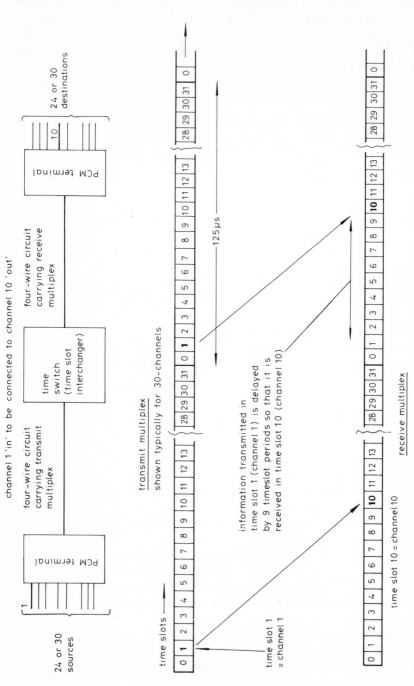

Fig. 2.8 *Time-division switching. The time-slot interchanger switches by delaying information received on a particular channel the appropriate amount for it to reappear on the required outgoing channel. The maximum delay occurs when e.g. channel 1 'in' is to be switched to channel 30 'out' and is just under 125 μs.*

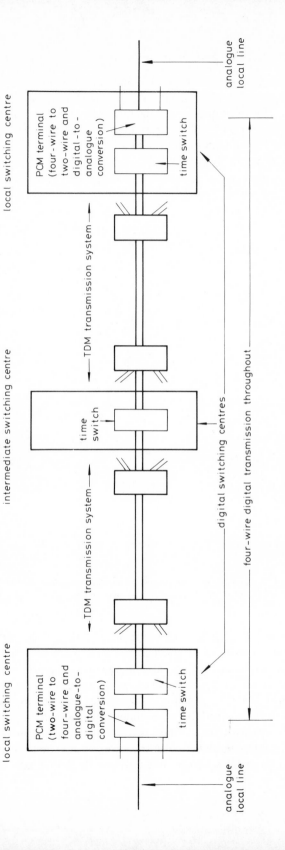

Fig. 2.9 *Integrated digital network.*

channels, circuit usage can be optimised. This approach, known as *common-channel signalling*, permits a much greater range of signalling information to be transmitted. Signalling information is transmitted on a packet-type data basis (see Sections 1.5 and 3.3), each packet carrying its own identity and destination information so that the signalling packets do not necessarily have to follow the same physical routes as the calls they control (Fig. 2.10).

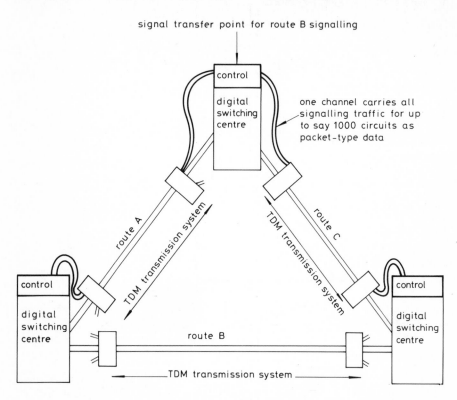

Fig. 2.10 *Common-channel signalling. One channel on each of routes A and C carries all signalling traffic for route B as well as for own routes.*

There are severe practical difficulties to the introduction of the IDN concept in developed countries with an established national space-switched network, since all existing switching centres and all analogue transmission systems must be replaced. Replacement on such a scale can only take place over a prolonged period; in the interim the old and new networks must coexist side-by-side with full interconnection maintained between them, to ensure that all users can continue to communicate with each other. Nevertheless, the established network operators in most major developed countries are already well advanced with such wholesale replacement schemes, not only for the immediate advantages already referred to but also for the future benefits discussed in Chapter 4. By contrast, the establishment of a completely new network permits the latest digital switching and transmission technology to be utilised without the penalty of large-scale replacement; thus the new network established in the United Kingdom by Mercury Communications, although not yet a national network in the true sense, is an IDN from the outset. The concept also offers considerable advantages for newly established private networks.

2.9 Mobile radio networks

Mobile radio networks, originally operator controlled and later fully automatic, effectively provide a radio extension to the local network. They link vehicle-mounted radio transmitter/receivers, known as mobile stations or *mobiles*, to one or more radio base stations which have direct access to the public telephone network. Some of the early systems, introduced from the 1950s onwards and mainly confined to large conurbations and the surrounding areas, are still operational. However, there is a shortage of available radio frequencies in the very high-frequency (VHF) band in which they operate, which limits their call-handling capacity to the extent that most networks have been unable to keep pace with demand.

To overcome the limitations imposed by the restricted radio spectrum available, new mobile radio network systems have been developed using a technique known as *cellular radio*. The cellular concept itself is not new but, as with so many other advances in telecommunications, implementation was not feasible until the advent of high speed microelectronic technology. The area to be served is split into a number of cells, which can most conveniently be shown diagramatically as hexagons. Each cell is served by its own base station, which has a transmitter limited in power output in order to reduce the possibility of reception beyond the cell boundary. Both transmitter and receiver must operate at different frequencies from those in the adjacent cells to avoid mutual interference. The cells are, therefore, arranged in a repeated group pattern so that the same frequencies can be reused in each group, beyond the normally expected range of interference (Fig. 2.11). This permits a limited range of frequencies to serve an unlimited area. The seven-cell grouping shown is a typical arrangement.

The central 'brain' of a cellular mobile radio network is located at one or more computer-controlled *mobile switching centres* (MSCs). Each MSC controls and is linked by direct circuits to a number of cells, and also has circuits to the public telephone network. The system is theoretically capable of handling any type of traffic appropriate to a telephone network (see Section 3.1). The MSC acts as a local switching centre but, in addition, has to perform other tasks peculiar to a cellular mobile network. These additional tasks arise from the fact that mobiles pass from cell to cell in the course of their travels. When a call is received from a mobile station, the required connection will be set up via the base station of the cell in which the mobile happens to be travelling at the time. For calls in the reverse direction, however, the mobile-network control computer needs to know in which cell the required mobile is located before a connection can be established. This is achieved by a process known as *registration*. Each mobile constantly monitors a base-station signal; as the vehicle

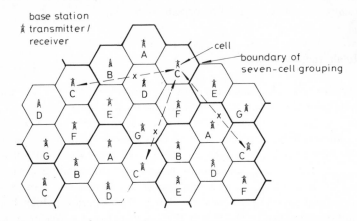

Fig. 2.11   *The cellular radio concept. The seven-cell grouping shown is typical. In this, seven different frequency bands are used, A to G. By dividing the area to be covered on a cellular basis, adjacent cells will always employ different frequencies, and the minimum distance x between base stations using the same frequency band ensures that mutual interference is rarely a problem.*

approaches a cell boundary, the signal transmitted by the home base station progressively weakens until a point is reached where the signal transmitted from the base station of the cell being entered predominates. At this point, the mobile reregisters with the control, thus letting the MSC know whenever it moves from one cell to another.

Each base station has available to it a number of radio channels for use by the mobile stations in that cell. The mobile stations are automatically switchable over all radio channels available to the network. When a radio connection is established with a moving vehicle to set up a telephone call, the channel allocated will be from those available to the cell in which the vehicle is currently travelling. Signal strength is constantly monitored and the crossing of a cell boundary is detected by comparing the relative strengths of the signals transmitted by the two base stations – the principle employed in registration. This will cause the call to be automatically switched to a new channel via the base station of the cell just entered (Fig. 2.12), a process known as *in-call handover* (known as handoff in the USA). The whole process of handover takes place within a few seconds, the only noticeable effect from the users' point of view being a break in transmission of less than half a second.

The very high radio frequencies used allow the mobile equipment and aerials to be more compact, and the equipment consumes less power since transmission is required over relatively short distances only, so that genuine hand-portable, battery-powered sets are a practical possibility. Cells vary from around 1 km across in city centres to many times this size in rural locations. It has to be accepted, however, that radio propagation conditions in city centres can be far from perfect, with tall buildings causing reflections and *shadowing*. Thus the siting of base station aerials is critical, and so-called *blind spots* leading to call failure are extremely difficult to avoid. For the same reasons, cell boundaries are 'grey' areas rather than the clearly defined demarcation lines depicted on a diagram or map, and systems must be designed to prevent a succession of handovers as a boundary is crossed.

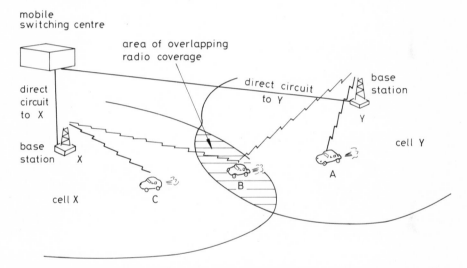

Fig. 2.12   *Cellular radio in-call handover. In position A a call is in progress using base station Y. In position B the mobile is crossing the boundary from cell Y to cell X. Signals will be monitored from base stations X and Y: when the signal from X predominates, the call will be automatically switched by the MSC from base station Y to base station X. The new radio channel allocated will then continue to be used as the mobile proceeds (position C) until the call is terminated or the mobile enters another cell.*

At present, a number of different system standards exist, all based on analogue techniques. The Nordic network which operates in Scandinavia is the only one that permits users to cross national boundaries. Two competing networks based on a common standard are rapidly approaching national coverage in the UK but in the USA, for example, separate networks are confined to individual cities and their environs. Work is in hand to establish a new European standard, probably employing digital techniques, which would allow the existing disparate networks within Europe to be replaced by a new common network. Mobile communication will undoubtedly become increasingly popular as wider international coverage and standardisation are achieved, but mobile radio services can be expected to remain an adjunct to the basic landline telecommunications networks.

Chapter 3

The transmission of data and other non-telephone traffic

3.1 Data, facsimile and video traffic on the telephone network

The method of communication used for public telephone networks, whereby independent point-to-point circuits are temporarily linked together by switching centres to establish a through connection between two users who wish to 'converse', is termed *circuit switching*. The calling and called users must both be available simultaneously for a call to be effective, but do not necessarily have to be human since automatic calling or answering devices may be employed. All circuits and switches involved in the connection must be held for the duration of the call. There is an initial set-up delay (which will worsen with increasing network loading) but, once set-up has been successfully completed, interaction is to all intents and purposes instantaneous and occurs at a transmission speed which is limited by the characteristics of the circuits involved.

The conversion of an established analogue public telephone network to an integrated digital network (IDN) is a lengthy and costly undertaking, involving the progressive replacement by digital equipment of all switching centres and transmission systems. Primarily intended for business users, digital transmission access to the IDN will be available as an alternative to analogue local-circuit access at extra cost (see Section 4.1), but it can only be provided where a connection can be established to a digital switching centre. Even though the amount of digital transmission and switching equipment increases each year, therefore, the network will be seen by most users as effectively an analogue transmission network since the normal access to it will still be via analogue local circuits. All traffic carried on the IDN will continue to be subject to the set-up and clear-down constraints imposed by a network designed to handle telephone calls, and must be capable of transmission over switched circuits of nominal 3 kHz bandwidth for analogue and 64 kbit/s bit rate for digital.

3.1.1 Data
Data traffic sensitive to set-up delay is generally unsuited to the telephone network. Nevertheless, there are still numerous applications where it is both

appropriate and economic to use the telephone network for data traffic. To understand the problems involved in transmitting data over telephone circuits, the nature of data signals must be examined in more detail. There are similarities between data transmission and computing, as might be expected since much of the data transmitted over telecommunications networks originates from or is destined for computers. As in the computing world, data signal pulses are considered in terms of two different states, designated 0 and 1, which are usually originated as different voltage levels rather than no-current and current conditions. A typical data signal might appear as in Fig. 3.1 – represented by 10001010. The speed of transmission is measured in bits per second (bit/s), where a bit is the smallest unit of a data signal – either a 0 or a 1.

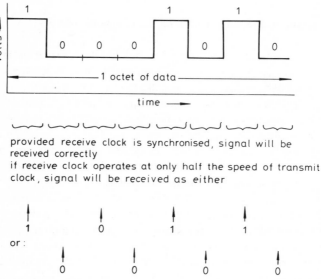

Fig. 3.1 *Data signal.*

It is essential that both transmitter and receiver are in step for meaningful data transfer; it will be seen from Fig. 3.1 that if the receiver was operating at, for example, half the speed of the transmitter, the signal could be read as either 0000 or 1011. To ensure that data signals are correctly interpreted, therefore, both transmitter and receiver must be controlled by a *clock* so that they operate at precisely the same speed. For relatively low speeds of transmission, each character transmitted is preceded by a start signal and followed by a stop signal; this concept, known as *asynchronous* or *start-stop transmission*, offers the simplest and cheapest solution to the problem. A number of very low-speed (telex-type) data channels can be accommodated on one telephone circuit by the use of frequency-division multiplexing, typically 12 channels operating at up to 110 bit/s each.

As the speed of transmission is increased, data signals suffer increasingly severe attenuation and distortion when transmitted directly over an analogue transmission local circuit. Conversion equipment is therefore necessary between each data terminal and its local circuit to change the digitally formatted data signals into line signals suitable for analogue transmission, and vice versa. The conversion is achieved by an electronic device known as a *modem* – a combined modulator and demodulator – which is interposed between data terminal and line. Modems use various modulation techniques to substitute audio-frequency tones for the digital bits, 0 and 1 being represented by different tones. A modem is designed for a specific operating speed, but can incorporate the facility to change to a fall-back lower speed where transmission conditions on an establish-ed connection are found to result in excessive errors at the normal operating speed.

By using modems, bothway data traffic at speeds up to about 2400 bit/s can be transmitted satisfactorily over an analogue telephone circuit, and higher speeds can be achieved by using two circuits simultaneously, one for each direction of transmission. For speeds above about 1200 bit/s, however, asyn-chronous transmission would introduce an excessive amount of errors. At these higher speeds, *synchronous transmission* is employed; periodic synchronising signals are transmitted with the data signals to ensure that receiver and trans-mitter remain in step (the process used in digital transmission systems). Never-theless, errors can still occur when, for example, poor transmission conditions are encountered on a telephone network connection. Unlike a packet-switched network dedicated solely to data traffic (Section 3.3), however, the telephone network is incapable of providing the facility automatically to detect such errors and invoke retransmission on a link-by-link basis.

A sequence of eight bits as shown in Fig. 3.1, each of which may be 0 or 1, is the basic building block of a telecommunications data signal; it is known as an *octet* (equivalent to a *byte* in computing terminology). The octets are not transmitted in a continuous stream but are conveniently segregated in sequential groups known as *blocks*, as in computer usage; in the case of packet-type data traffic, the blocks are assembled into *packets* and *frames* (see Section 3.3). There are 256 different possible ways of combining 0 and 1 in groups of eight; thus an octet can represent up to 256 different characters, depending on the code used. Typical of the more recent data transmission codes is the CCITT International Alphabet number 5 (IA5), designed for synchronous transmission and virtually identical to the American Standard Code for Information Interchange (ASCII), in which seven of the eight bits are used to represent data information.

A telephone is normally used to set up a connection for a data call over the telephone network, the distant end acknowledging by telephone before data is transmitted over the line via modems at each end. An auto-answer modem option allows the connection to be established without human intervention at the called end, which might, for example, be a time-sharing computer bureau. Similarly, an auto-dial facility permits the transmitting data terminal automatically to seize the line and transmit the required number to set up a call. Data calls can be originated or received via a cellular mobile radio network but, to avoid excessive error rates, specially designed terminals are essential for use on the move because of the problems posed by in-call handover and variations in transmission quality over radio channels (Section 2.9).

Where a digital local circuit is available it becomes possible to input data signals without the use of a modem and, by using time-division multiplexing, various combinations of data transmission paths can be provided within the 64 bit/s capacity of a single circuit. Data transmitted directly over a digital circuit is less prone to error than over an analogue line, and asynchronous transmission at up to 19.2 kbit/s is possible. The concepts are described in Section 4.1.

3.1.2 Facsimile

Fascimile, popularly known as 'fax', is the oldest form of electronic mail, having been conceived more than a century ago. Modern facsimile terminals are self-contained devices which permit the remote 'photocopying' of a sheet of paper on which information has been written, typed, drawn or photographically developed. The page of information to be transmitted is systematically scanned by a light-sensitive electronic device to detect the variations between black and white along the scanning path. The information thus obtained is converted into a data signal which can be transmitted over the telephone line. A repetition of the scanning path at the receiving terminal allows the signal to be reproduced on paper as gradations between black and white – in effect, as a photocopy of the original document. A call is set up over the telephone network in the same way as for a data call, using an associated telephone, a keypad built into the machine or the auto-dial facility if available. An auto-answer facility is also obtainable on the latest machines.

Facsimile terminals to the current CCITT group 3 standard use digital data transmission techniques to send information at 4.8 kbit/s over analogue local circuits, offering a copying time of about 1 minute per A4 sheet. Where a digital local circuit is available (Section 4.1), the latest group 4 standard terminals permit faster transmitting speeds, reducing the copying time to a few seconds.

3.1.3 Video

The transmission of video (moving picture) signals requires a huge frequency bandwidth, typically 5.5 MHz for analogue colour television signals. It is impossible to transmit such signals over a switched public telephone network, and the wideband leased 'circuits' used take up the frequency spectrum of transmission systems that would otherwise be occupied by many hundreds of telephone circuits. Typical of current video telecommunications services are video conferencing and remote video alarms; both require a video camera to record the scene being observed. The camera scans the image seen through its lens just as a facsimile terminal scans the document to be copied, detecting variations of light and shade along the scanning path which are then converted to electronic signals. These signals, received over a suitable transmission path, are used as in television reception to reconstruct the scene on a monitor screen by means of a synchronised repeat of the scanning process. For colour transmission, again as in television, three primary colours are detected separately to produce separate signals for each colour. Video conferencing requires two separate wideband circuits, one for each direction of video transmission, plus the facility to transmit speech. An alarm circuit requires video transmission in one direction only and sound is unlikely to be necessary.

Digital transmission techniques are not used by broadcast television transmitters, but digitally encoded video signals offer advantages for transmission over telecommunications circuits. Elaborate digital coding techniques can be used to eliminate redundant information (for example, information from the picture background which does not change from scan to scan). Alternatively, the amount of data to be transmitted can be reduced by using slow-speed scanning. The selective coding approach is particularly suited to video conferencing, where a number of people sit in a fixed setting and only their movements need to be transmitted; advanced coding processes may also incorporate speech to avoid the need for a separate speech circuit. The slow-scanning approach produces a series of 'frozen' images with no attempt at continuity, and is appropriate for remote surveillance services. These techniques open up the possibility of transmitting acceptable slow-scan black-and-white pictures as data signals over a 64 kbit/s switched digital connection; however, even with selective coding, continuous colour video still requires many times this capacity (see Section 4.1 for possible future developments).

3.2 Telex networks and teletex

Most present-day public telex networks are self-contained entities, linked internationally, separate from the telephone network but carried on the same transmission network. When telex services were first established some 50 years ago, however, it was the normal practice to share the circuits of the telephone network, since both services had a common need for circuits to be connected for the duration of a call and then released. It was then the practice to use an associated telephone to set up calls, in the same way that data calls may be set up over the telephone network today. With the growth of telex traffic, however, the limitations imposed by the telephone network made it advantageous to establish a separate telex network with independent traffic routes and switching centres. 'Dial-up' access is provided, but inter-switching-centre signalling is not required. Instead, the signalling functions are performed by an exchange of messages: for example, each customer has a unique answer-back code which is automatically returned to a calling terminal to provide confirmation of identity before transmission commences. Telex is exclusively a business service, attractive internationally because of its immunity from the effects of time zone differences, since attendance by the called party is not required outside the normal local business hours. A very high proportion of telex traffic is, therefore, international.

The telex service, by virtue of its early dependence on the telephone network, was initially circuit switched, and this concept was retained after network separation because telex switching centres followed the then current telephone-type electromechanical practice. A 'call' might consist of a number of messages in each direction. The availability of electronic switching centres having store-and-forward facilities has allowed an entirely different method of communication between switching centres to be introduced, known as *message switching*. Instead of two users exchanging messages in both directions within the 'conversation' time between the set-up and clear-down of a single call, each message is, in effect, treated as a separate call. The destination address, added to each message, is used by each switching centre to select an appropriate free circuit for onward transmission. Where there is no free circuit or the called user's terminal is engaged, the switching centre will store the message and make periodic attempts to transmit it until success is eventually achieved. Messages can also be accepted for onward transmission at a predetermined time. The message-switching store-and-forward concept requires every switching centre in a previously circuit-switched network to be replaced, a task which can only be accomplished progressively over several years.

Transit time over a circuit-switched telephone network is unaffected by congestion once a call has been established; instead, set-up is delayed or prevented altogether when the network is overloaded. By comparison, congestion in a store-and-forward telex network will increase switching centre *message handling times* and thus the overall *transit time*, but this will not be apparent to the sender as long as the network continues to accept and store messages; the storage capacity effectively enables peak loads to be accommodated with less circuit capacity. The sender will only become aware of a problem when, exceptionally, a message cannot be accepted because the storage capacity is full.

The speed of telex operation has always been measured in terms of the *modulation rate* – in other words, the speed at which signals are transmitted along a line. The unit used is the *baud*, which dates from the early days of telegraphy. A baud is equal to one signal element per second. Where each signal element represents one data bit, 1 baud = 1 bit per second). The modulation rate in bauds does not necessarily have to be the same as the rate of data transfer in bits per second, however, If, for example, four signal levels are used instead of two, then two data bits can be represented by one signal element; the data transfer rate is doubled for the same modulation rate and 1 baud = 2 bits per second (Fig. 3.2). This idea is widely used in data communications generally, with as many as 32 signal levels being employed to speed the rate of transferring information without increasing the line transmission speed.

Electromechanical teleprinters operate at 50 bauds, their speed being limited by the printer and the direct-current signalling method used. Electronic telex terminals have faster printers and employ the voice-frequency signalling principle used in modems for data transmission over the telephone network, operating at speeds up to 300 bauds. Modern message-switching store-and-forward telex switching centres make use of the stored-program control and digital switching techniques which are a feature of the latest telephone-network switching centres, and provide automatic speed conversion over the range 50 to 300 bauds so that two terminals can exchange messages whilst operating at different speeds.

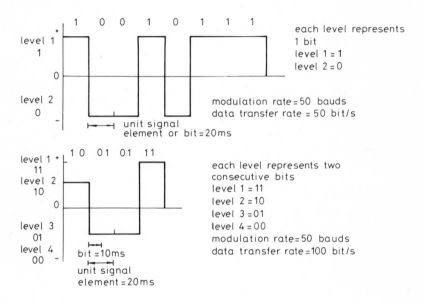

Fig. 3.2 *Modulation rate and data transfer rate.*

The standard CCITT International Alphabet number 2 (IA2) used for telex employs seven signal elements per character and inherits the old telegraphy terms *mark* and *space* instead of 1 and 0. Transmission is asynchronous; hence the first of the seven signal elements transmitted is always a space (the start signal) and the last is always a mark (the stop signal), and only the five middle signal elements represent the character being transmitted. This limits the number of possible characters, but the typewriter-style keyboard allows the same signal to be used for either a letter or a number, depending on whether it has been preceded by the signal transmitted as a result of depressing the letters or numbers shift key.

Electronic telex terminals offer more facilities than electromechanical teleprinters and can be integrated into an office information system. Auto-dialling is a standard feature, and they provide electronic storage and display facilities for the limited word processing of messages for transmission. The storage capacity also permits transmission on a store-and-forward basis with automatic repeat attempts, and can be used to hold messages received via the telex link or input from other data terminals in the office. These terminals can be connected to the telex terminal by synchronous links using IA5 and operating at much higher speeds than the telex link – typically up to 9.6 kbit/s.

Teletex (which should not be confused with the television information service known as *teletext*) has been defined internationally by CCITT as a new text communication service to offer improved facilities over the existing telex service. Setting up, transmission and reception is fully automatic and provision is made for end-to-end data error detection and retransmission (see Section 3.3 for explanation). Although specially designed terminals are available, these are not essential; users can employ an adaptor to make existing memory typewriters, word processors or computer terminals compatible with teletex. Teletex access, therefore, can be fully integrated with office information systems. Synchronous transmission at 2400 bit/s is employed, with IA5 instead of IA2 as the data signalling code (see Section 3.1.1). A page of text can be transmitted in about 20 seconds, much faster than telex, and the code enables teletex to support a much wider range of characters. Teletex does not require a separate network but makes use of the circuit-switched telephone network for interconnection via a normal telephone local circuit; large-scale users can obtain more economic rates by accessing the packet-switched network from their telephone lines (Section 3.3). Although interconnection with the telex network is provided so that teletex users can exchange messages with telex users, teletex can be expected to eventually supersede telex.

3.3 Packet-switched data networks

The message-switching store-and-forward telex network was designed primarily to enable *people* to exchange messages in text form. People recognise a complete message as the unit of information and expect there to be no restriction on message length (other than that imposed by the cost of transmission). A message is, therefore, accepted and delivered by the telex network in its entirety. The accent is on reliability rather than speed, and rapid interaction between users is not possible. These limitations of a message-switching network make it totally unsuited to the interactive exchange of information between a data terminal and a computer, and it was to meet this need that packet switching was developed. *Packet switching* makes use of the principles of message switching and store-and-forward working, but data messages are transmitted as a number of equal-length packets; the concept of packetised data was introduced in Section 1.9. Public packet-switched networks are linked internationally in the same way as telex networks, and interconnection facilities enable data users on packet-switched, telex and telephone networks to communicate with each other.

Packet-switching networks are designed for rapid response and handle information in much the same way as a computer. The fixed, short packet length avoids the prolonged holding of circuits experienced with message switching. Storage queues at switching centres can, therefore, be kept short and the total amount of storage in a packet-switching network is low compared with a message-switching network. The concept of speed matching as used in the telex network is further developed to encompass a much wider range of transmission speeds. A packet-switching network can be compared with a time-sharing bureau computer, which is able to serve many users apparently simultaneously because it is much faster in operation than any of them. Whereas the computer stores information internally, network storage is distributed among the switching centres. By switching rapidly between users and employing this storage capacity as a *buffer*, to smooth traffic flow and permit speed matching, both computer and network offer users a very rapid response, accepting or transmitting data at a rate appropriate to the individual user. *Cross-network delay* for a public packet-switched network, dependent on switching-centre *packet handling times*, is typically about one-third of a second.

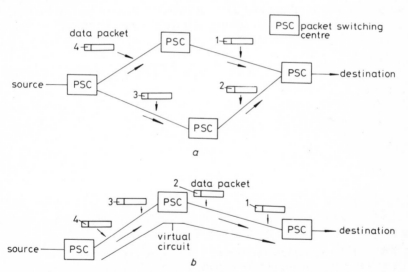

Fig. 3.3 *(a) Datagram working. Packets routed independently of each other: they may follow different routes and arrive out of sequence. Reassembled in correct sequence at destination. (b) Virtual-call working. Initial route established by an exchange of call set-up packets between source and destination. Time share on this route is then maintained for transmission of subsequent data packets until source signals end of transmission. Packets arrive at destination in correct sequence.*

The simplest form of packet-switched network operation, in which each packet is dealt with as a completely independent self-routing call in its own right, is known as *datagram* working (Fig. 3.3a). This mode of operation is commonly used in private networks, but in public networks a variant known as *virtual-call* working is preferred (Fig. 3.3*b*). This takes account of the fact that many users need to exchange large amounts of consecutive data. Some of the dynamic properties of datagram working are removed by ensuring that, once a connection between two users has been established, their transaction will continue to employ the same routing until it is completed; in effect they are allocated an assured time share of a connection, termed a *virtual circuit* because it is in effect the same as a switched-circuit connection. This simplifies matters by ensuring that the packets from any one user arrive at their destination in sequence.

Communication between computers involves the transfer of blocks of data, each block containing a number of sequential data octets (see Section 3.1.1). Data blocks are grouped together to form packets, each packet being automatically preceded by additional data, termed the *header*, which contains information to identify the packet and its destination. A *packet* typically contains up to 16 blocks, that is, 128 octets of user data, plus a 3-octet header (Fig. 3.4a). This process of packetisation is not a network function; it is carried out within the user's data terminal equipment (then described as a *packet-mode terminal*) or by an add-on *packet assembler/disassembler* (PAD). Both transmit synchronously over the network using the IA5 data signalling code, but the PAD can accept synchronous or asynchronous data signals from a user terminal.

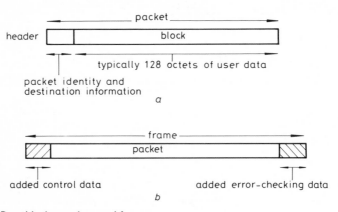

Fig. 3.4 *Data blocks, packets and frames:*
(a) Blocks into packets (packetisation)
(b) Packets into frames (for transmission over network).

Imperfect network transmission conditions can result in a received signal which does not precisely match that originally transmitted. This rarely matters with speech, but data signals can be distorted to the point where their meaning is lost. Packet switching provides a means to correct such errors on a link-by-link basis within the network. Extra data octets are added to the front and rear of each packet before transmission, to facilitate both control and error correction; the enlarged packet is then known as a *frame* (Fig. 3.4b). The error-checking octet within each frame is automatically inspected at the receiving end of each circuit; correct receipt is positively acknowledged to the transmitter but an error causes a negative acknowledgment to initiate a retransmission of the frame concerned.

Access to a public packet-switched network may be (a) direct via a digital circuit or (b) via a telephone-network/packet-network interface over a normal analogue telephone circuit (Fig. 3.5). With method (b), signalling speed is restricted (Section 3.1.1), errors introduced by the telephone circuit cannot be detected, and a separate circuit is required for each data-terminal/telephone-network connection. The direct-access method (a) permits data to and from several terminals to be handled simultaneously on one circuit. This is achieved

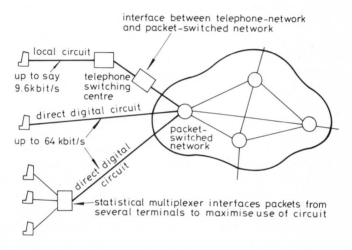

Fig. 3.5 *Packet-network access.*

by a process of interposing the packets from different sources, known as *statistical time-division multiplexing* or simply statistical multiplexing. Statistical multiplexing is a dynamic process. Packets are assembled in a queue ready for transmission to the network, their place in the queue being entirely random as regards source and determined solely by timing. Packets are transmitted in the order they arrive in the queue, in the form of frames with control and error-checking data added. In the reverse direction, frames received from the network are stripped of their control and error-checking data and, as packets, queued in a similarly random order of destination. The header information is interrogated to determine packet destination so that all packets making up a complete message may be directed in sequence to the required terminal. The receiving user will be unaware of this process since the message will arrive at the terminal as a continuous data stream, stripped of all headers. Transmission over a direct-access circuit may be at speeds up to 64 kbit/s; the higher the speed, the greater the volume of data traffic that can be handled.

Packet-network switching centres are generally interconnected by 64 kbit/s circuits, packets being interleaved by statistical multiplexing for transmission over each circuit. Error correction is carried out between switching centres on a similar basis to that over the local access circuit. The added control and error-checking information is stripped from the packets on entry to the switching centre and the packets are then digitally switched to the required route, determined by interrogating each header. The packets are again supplemented by control and error-checking data before retransmission as frames, and the data momentarily remains in store after transmission until its correct receipt at the distant end has been acknowledged. Routing does not have to be confined to a particular group of circuits as in a circuit-switched network; the first available time slot on one of several possible routes may be selected, a procedure known as *alternative routing* (see also Sections 7.5 and 8.1). As previously noted, with virtual-call working the same routing is retained for the duration of a transaction between two users, whereas with datagram working all packets are routed independently.

3.4 Data interfaces, protocols and open systems interconnection

Data communication requires machines to interact directly with each other, without human intervention. Machines are incapable of discretion or intuition and the way in which they interact must be very precisely defined. It has already been shown that data terminals and networks operate at widely different speeds and may transmit data signals synchronously or asynchronously, possibly using different signalling codes. These are the visible characteristics; beneath the surface, as it were, information is exchanged in the form of precisely structured blocks, packets and frames of data, and communication requires precise messages to be interchanged for control and error-checking purposes. Each terminal and network termination, in fact, presents its own particular set of requirements to other terminals or networks, which must be matched if data signals are to be exchanged between them. These requirements are defined as a number of sets of rules which embrace every possible factor that needs to be taken into account to ensure effective communication.

A set of rules which defines the conditions necessary for the exchange of data between two points is termed a *protocol*. Protocols are organised on a number of different layers as in the example shown in Fig. 3.6, covering the exchange of data between a data terminal and a data switching centre. This layered approach means that requirements for different aspects of the conditions necessary to permit the transfer of data between two users can be specified separately; each layer is dependent on the layer below it and the conditions demanded by all layers must be satisfied for successful data transfer to be possible. Only at the physical layer is an actual physical connection established, for example, the plug-and-socket connection between a data terminal and a modem; the relationship between two entities at the higher layers is purely conceptual. The boundary between two physical or conceptual entities is termed an *interface*. Some national protocol standards have gained widespread acceptance, notably those published by the US Institute of Electrical and Electronics Engineers (IEEE), and the internationally agreed standards are set out in a series of CCITT recommendations; the V series relates to operation over analogue transmission networks, the X series to digital networks and the I series to the integrated-services digital network (see Section 4.1). The most commonly used of these are included in the glossary at the end of the book.

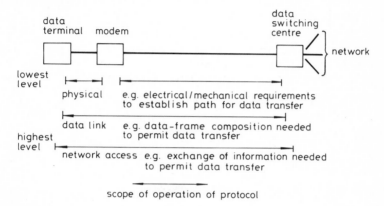

Fig. 3.6 *Protocol layers.*

Faced with the rapid growth in demand for private data-communications networks, some major manufacturers have developed their own protocol standards in order to establish a proprietary network architecture. (This has an immediate advantage for the company concerned, ensuring a closed market for its products once networks have been established, but in the longer term competing manufacturers often produce compatible equipment that is cheaper and/or more efficient.) The situation has resulted in a proliferation of data communications equipment standards, quite separate from the standards established by CCITT for data communication over public telecommunications networks. This clearly limits the scope for data users to communicate with each other and has led to the concept of a single integrated set of standards under the heading of *open systems interconnection* (OSI). OSI is sponsored by the International Standards Organization (ISO) with the close collaboration of CCITT and supported by manufacturers and network operators worldwide.

The aim is that any data communications system implemented to the OSI standards should be 'open', that is, able to communicate with any other system implemented to the standards throughout the world. Full OSI is a long-term goal which can only be achieved by taking existing protocols into account in the interim. It is being pursued through the establishment of a framework of standards based upon the *OSI basic reference model* (IS7498 refers), which extends the layered concept referred to earlier. Work on the OSI reference model began in 1978. Discussion to establish a full range of standards to meet all possible situations is a continuing process, having expanded beyond the original idea of communication between computers to embrace the public telecommunications networks over which that communication may take place.

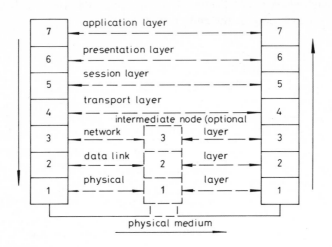

Fig. 3.7 *The OSI basic reference model*

The OSI reference model is based on an abstract structure of seven hierarchical layers (Fig. 3.7). Each layer is defined in terms of the visible external behaviour required of a system, allowing manufacturers the freedom to implement systems in their own way providing these external requirements are met. Two different types of standard apply to each layer; one defines the *service* to be provided by the layer, relating the service available from the layer below to the service to be offered to the layer above, and the other specifies the *protocol(s)* needed for two systems to communicate within that layer to provide the defined service. Provision is made for one or more intermediate *nodes* to act as protocol *converters* where communication requires the use of different protocols over different stages of a connection between two data users, or where no common protocol is currently available. Progress through the reference model takes place from layer 7 to layer 1 for transmission and in the reverse direction for reception.

Below layer 1 is the *physical medium* – the transmission medium, either cable (metallic or optical fibre) or the ether (radio). Several different transmission media may be encountered over the path of a communication between two data users. The requirements for data transfer are distributed between the seven layers. Layers 1 to 4 (sometimes collectively referred to as the *transport service* layers) are concerned with interconnection over a network and are therefore of concern to the network operator. Layers 5 to 7 are concerned with communication between the data users. The requirements covered by each layer are complex to explain fully, but in essence are as follows:

Layer 1: physical   Electrical/optical and mechanical requirements to permit a continuous path for data transfer to be provided over the physical medium.

Layer 2: data link   Data-frame composition including header, control and error detection/correction requirements, to enable data transfer over the path provided by layer 1.

Layer 3: network   Interconnection of the paths provided by layer 1 carrying the data composed by layer 2, including the requirement to deliver the data in the correct sequence.

Layer 4: transport   Matching between data users of any multiplexing requirements as determined by, for example, the teletex, telex or packet-switched methods of communication.

Layer 5: session   Matching between data users of data-transfer modes (two-way simultaneous or alternate, or one-way only), synchronisation requirements, etc.

Layer 6: presentation   Matching of data structures and coding method between data users.

Layer 7: application   Included for completeness but not directly appropriate to this book since it concerns the relationship between a data user's computer program and the communications environment.

Chapter 4

The integration of public networks and services

4.1 Local-network developments leading to the integrated-services digital network

In Chapter 2 the development of an established public telephone network was followed to the present, when digital transmission systems and switching centres are replacing the earlier analogue equipment to create an integrated digital network (IDN). This provides telecommunications users with a 64 kbit/s path between local switching centres, analogue-to-digital conversion occurring at the point where the user's analogue local circuit enters the digital switching centre. It will take many years to achieve a complete nationwide IDN on this basis. However, there are obvious advantages to the business community in establishing an early core IDN to link the more important business centres; this core network can then be progressively expanded whilst coexisting with the residual declining analogue network.

Clearly, a user would derive maximum benefit from the IDN if, in addition, the local circuit could support digital working at 64 kbit/s. Users so equipped would then be able to communicate by means of speech, data, facsimile or restricted-bit-rate video signals over a 64 kbit/s path, end-to-end. The problem is that a single local-network cable pair is not, by itself, capable of supporting simultaneous bothway digital transmission because of the impossibility of separating the signals in each direction. There is also the question of the increased loss incurred by higher-speed digital signals; this limits the distance over which transmission is possible. A separate pair for each direction of transmission as used for higher-speed data transmission (Section 3.1.1) would increase the achievable bit rate, but at the expense of a greatly increased local-network cable requirement. However, two new transmission techniques have now been developed to overcome the limitations of a single cable pair.

The *burst-mode* technique solves the problem by sending the digital information over the local circuit in short bursts, alternately in one direction and then the other. Each burst of a fixed, small number of bits is transmitted at a much higher rate than its original 64 kbit/s and as a result is compressed in time (Fig.

4.1a). In effect, the time occupied by a burst of bits is halved before transmission, and the space thus made in the 'real' timescale allows for a similarly compressed burst of bits to be sent in the opposite direction. The circuit is not

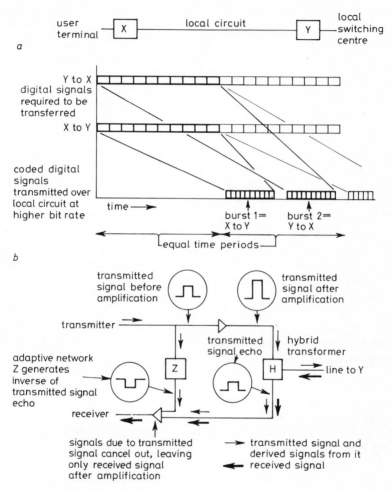

Fig. 4.1   *Digital transmission over a local circuit:*
(a) Burst mode
(b) Echo cancelling: equipment at X

called upon to carry signals in both directions simultaneously; hence the difficulty of separating the two sets of signals does not arise. The increased transmission loss which results from transmission rates in excess of 64 kbit/s restricts effective operation to a distance of a few kilometres, but this is sufficient to meet the needs of most users.

The other technique, *echo cancelling*, uses more complex circuitry to attack the heart of the problem. Essentially, the method relies on the fact that the difficulty of simultaneous high-speed digital transmission in both directions arises from the unavoidable occurrence on a circuit of spurious transmitted-signal 'echoes'. These interfere with the wanted signals being received in the opposite direction, already weakened by transmission loss, and prevent their separation and accurate detection. By generating counterbalancing signals to cancel out these echoes, echo-cancelling equipment makes simultaneous both-way transmission possible without the need, inherent with the burst-mode approach, to increase the transmission rate (Fig. 4.1b). This permits acceptable transmission over slightly greater distances than with the burst-mode technique.

These developments enable a local circuit to give *integrated digital access* to the telex and packet-switched networks as well as the basically telephone IDN, and the complete network is termed an *integrated-services digital network* (ISDN) (Fig. 4.2). The ISDN can only be introduced in step with the IDN since digital working is dependent on there being an accessible digital switching centre. The declining residual analogue telephone network will, of course, remain until the last of the analogue equipment has been replaced. The ISDN concept has been heralded as a major breakthrough in the history of telecommunications. It provides complete integration of all services over one circuit, as seen by the user, and enables the network operator to offer a wider range of non-telephony services. These can be provided more cheaply because of the economies of scale offered by a network optimised for the primary purpose of carrying bulk telephone traffic. Data traffic, in particular, benefits from significantly reduced call set-up times (a second or two instead of as much as 25 seconds) and the possibility of increased transmission speeds (up to 64 kbit/s). An ISDN user has the option of making:

(a) A telephone call to any other telephone user: a wholly digital path is not necessary and the call may be routed partly over the residual analogue telephone network.
(b) A data-only call to any other compatible data user, whether on the IDN, telex, packet-switched or residual analogue telephone networks.
(c) A voice-plus-data call via the IDN which allows users to switch between voice and data in mid-call: in this context, data embraces all types of non-telephony traffic carried over the telephone, telex and packet-switched networks.

Both burst-mode and echo-cancelling transmission methods require additional equipment at each end of the local circuit. At the user end, the digital encoding/decoding (for analogue telephones), bit-rate adaption (for data and facsimile terminals), protocol conversion (for data terminals), transmission and signalling functions are all carried out by *network terminating equipment* (NTE) inserted between the user's terminal(s) and line (Fig. 4.3). Signalling by the normal direct-current means is no longer possible and an additional lower-speed digital

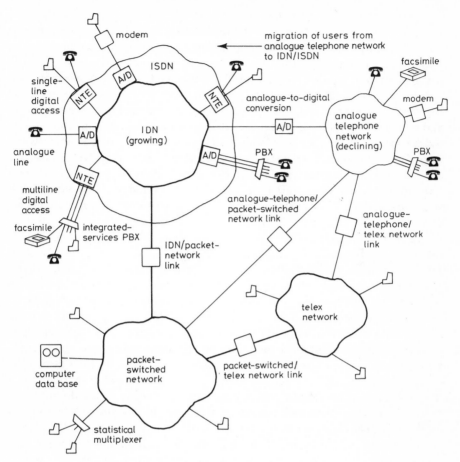

Fig. 4.2 *The integrated-services digital network. NTE is network terminating equipment: see Fig. 4.3*

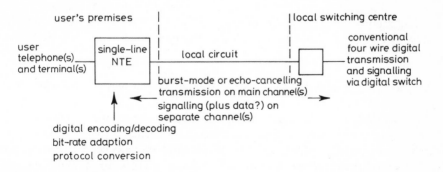

Fig. 4.3 *Network terminating equipment*

channel is multiplexed over the same pair for this purpose, the data-message-type signals used offering a greatly enhanced, faster signalling capability. At the local switching centre, another terminating unit converts between the line signals and conventional four-wire digital transmission and signalling. This terminating unit takes the place of the analogue-to-digital conversion and direct-current signalling equipment previously required. The CCITT standard for single-line integrated digital access provides for two 64 kbit/s main channels and a 16 kbit/s signalling and packet-switched network access channel, giving an overall transmission rate of 144 kbit/s. However, some established network operators in the process of converting to an IDN – including British Telecom (BT) in the UK – introduced the ISDN concept initially on a pilot basis, using interim equipment developed in advance of this standard. The interim BT single-line standard, for example, provides for one 64 kbit/s main channel and two 8 kbit/s signalling/data channels.

The user requiring multiline access – for example, to serve a PBX – is catered for by the installation of a PCM transmission system adapted to provide a number of main and signalling/data channels over two local-circuit pairs. However, the established local network was not designed for transmission at the high bit rate demanded by a PCM system, and some cable pairs will prove to be unsuited to the task. Where a PCM system would require the provision of a new metallic-pair cable, the installation of an optical-fibre cable with its attendant transmission equipment may offer an economic alternative. Some cable pairs will also prove to be unsuitable for transmission at the bit rates demanded by echo-cancelling and burst-mode equipment. Situations will arise where there is no spare or usable capacity on existing cables, or where cables do not exist; the last-named problem will always be faced by the operator attempting to establish a new network, as is Mercury Communications in the UK. Low-powered radio transmission systems can then offer an economic alternative and digital microwave radio systems have been developed for the local network to meet this need. Small *point-to-point* radio systems provide the required channels between two locations, using similar techniques to the larger systems employed on main transmission routes, and *multipoint* radio systems employ a central station with the transmitting/receiving aerial typically giving coverage over a 90° or 120° sector to serve a number of individual users within that sector. A suitable aerial is, of course, required at each user location.

4.2 Integrated speech and data: the service-independent network concept

With the growth of the teletex service – based on the telephone network and thus on the IDN – there is likely to be a declining demand for the telex service. Indications are that a separate telex network will eventually cease to be economically viable; the ISDN can then be expected to give access to the IDN and only one other network, the packet-switched network (to which direct connection will continue to be provided where required). In spite of the improvements in set-up time achieved with the introduction of digital access, the IDN will remain a circuit-switched network optimised for bulk telephone traffic and the packet-switched network will continue to offer a preferable alternative for some types of data traffic. The ISDN will offer a range of speech, data, facsimile and video services capable of transmission over a 64 kbit/s path, but switched high-quality video communication will continue to be excluded unless there is a technology breakthrough (as yet unforeseen) to reduce the very high transmission speed currently required.

The future of public telecommunications networks beyond the ISDN must be largely a matter for conjecture but, from the technological point of view, there is only one way in which development can proceed – towards a network this offers communication paths capable of transmitting even more information more quickly than at present. The possible end results are higher-quality speech, higher-speed data, faster and better-quality facsimile, and video pictures of television quality. The accent must be on improved standards to match increased customer expectations in every sphere of telecommunications, at lower real cost to the network operators and thus to the customers. How might these improved standards be achieved?

The answers to this question are already being explored in research centres throughout the world, with a view to establishing a pattern for telecommunications network development into the next century. Essentially, for improved standards to be achieved, a circuit must be capable of carrying information at a higher bit rate. Overall transmission capacity poses no problem with the availability of optical-fibre transmission systems. Capacity is now allocated on the basis of 64 kbit/s channels, but multiplexing already provides for non-switched circuits operating at transmission speeds in excess of 64 kbit/s where required. There would be no technical difficulty in designing new system terminal equipment with a different basis of allocating bit-rate capacity. Digital switching systems now being installed for the IDN are, however, capable only of switching fixed-speed 64 kbit/s paths on a circuit-switched basis; any move towards the transmission of more information more quickly requires a change to this concept. It is, therefore, the long-term future of switching that is occupying the attention of scientists and engineers.

The average residential telecommunications user's requirements will be adequately catered for by the current 64 kbit/s bit-rate standard for many years to come. Among the improved standards which business users might press for, high-quality video poses the major technical switching problem; it is likely to be needed by only a minority of users, yet requires transmission speeds of many times the 64 kbit/s bit rate. In a fixed-transmission-rate, circuit-switched network, the switched paths must cater for the highest bit rate required. It would clearly not be economic, or indeed practicable, to provide switched circuits on the basis that all telecommunications users might require high-quality video transmission. The answer, therefore, almost certainly lies in some alternative switched-network concept which would allow the bit rate to be dynamically varied to suit a particular user's needs. This has been termed *variable-bit-rate switching*, and might possibly employ ideas first developed for data traffic. For example, all information including speech could be transmitted and switched as data packets, using a switch architecture based on the private local-area-network ring (introduced in Section 5.3). It is the bread-and-butter telephone conversation that poses the chief technical problem with this concept: because of the need for immediacy and continuity, speech information would need to be given priority over all other traffic on the network. Such a switching system is already a theoretical possibility and could become a practical reality within a decade.

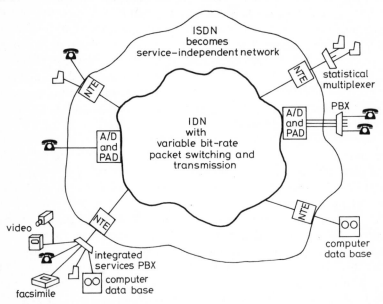

Fig. 4.4   *A service-independent network of the future. NTEs incorporate packet assembly/ disassembly (PAD) function where this is not provided by PBX or user terminals*

   The progressive introduction of variable-bit-rate switching principles to the IDN, coupled with (a) changes to the way in which bit-rate capacity is allocated by transmission systems and (b) further advances in local-network transmission technology to permit matching higher operating speeds over local circuits where required, could eventually result in a single integrated network able to carry all types of traffic – a truly *service-independent network* (Fig. 4.4). This would allow the separate packet-switched data network to be dispensed with. By allocating (and varying) both the capacity and transmission speed needed by each user as required at any particular time, the need for network resources would be minimised, making possible a more highly efficient network than ever before.

However, any change from the concept of a circuit-switched network will require network operators to embark on another wholesale and very costly replacement programme; even if the decision to go ahead were to be taken now, both economics and the time needed for system development are unlikely to allow a start to be made before the mid 1990s.

Chapter 5

Private networks

5.1 Leased private circuits on public networks

The precise distinction between public and private telecommunications networks is debatable in countries such as the UK and the USA, where there are privately owned public networks. However, the generally accepted understanding of a private network is that it is primarily intended to carry internal traffic originated by, and destined for, employees of the organisation it exists to serve. Most private networks also have access to public telephone and data services. These networks range in size from those which provide service to a single site to those which link many sites nationwide or even internationally. A small PBX and its extensions is an example of a single-site private network in its simplest form; by contrast, a multisite private network may consist of many interlinked switching centres, carrying both speech and data traffic.

The owner of a network which links a number of scattered sites is rarely able to interconnect those sites by cable; if direct microwave radio communication is not feasible, therefore, circuits may be leased from another – usually public – network or a privately operated satellite communications system. Satellite systems can offer an economic alternative for communication over long distances, but most intra-national private networks use circuits leased from a public network.

Leasing at the simplest level takes the form of an individual point-to-point circuit to public-network standards. A leased private circuit of this type offers the customer 3 kHz bandwidth for analogue transmission or 64 kbit/s bit rate for digital transmission, or is marketed specifically for data transmission at a particular bit rate – typically 2.4, 4.8 or 9.6 kbit/s for analogue, and 19.2, 48 or 56 kbit/s or multiplexed combinations of these and/or the lower speeds for digital transmission. The complete circuit is made up from two local circuits, one at each end, linked over long routes by channels extracted from a transmission system or from several transmission systems in tandem (Fig. 5.1). For end-to-end digital operation all sections, including the local circuits at both ends, must operate digitally; the local circuits are usually provided over two

Private networks 59

cable pairs, one for each direction of transmission, to avoid the need for the more complex single-pair methods described in Section 4.1 for the ISDN. Analogue local circuits for the higher data transmission speeds are also provided over two cable pairs.

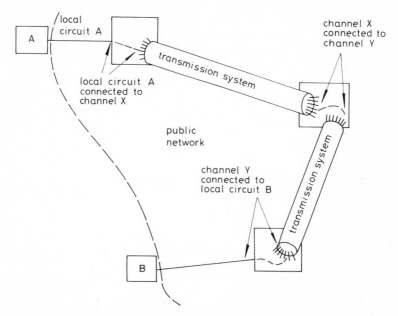

Fig. 5.1 *Leased point-to-point private circuit, linking customer's premises A and B*

The local circuits and transmission-system channels used for a leased private circuit are normally directly joined in tandem, any switching centres *en route* being bypassed. However, the low-cost time switching used in digital public-network switching centres permits a leased circuit to be routed *through* a switching centre instead of bypassing it, using a 'semi-permanently' switched connection. Each switching centre *en route* is programmed to 'link' the time slots of the required local circuits and/or transmission-system channels for a predetermined period, instead of on the per-call basis used for switching public-network calls (Fig. 5.2: compare this with the virtual-circuit packet switching of data, Section 3.3).

An alternative and equally flexible approach to the provision of digital private circuits utilises capacity on a public digital transmission network to establish a 'network' of channels which are reserved exclusively for private circuits. The 64 kbit/s channels are multiplexed as standard 2 Mbit/s PCM streams. Interconnection of the required channels, on a semi-permanent basis, is achieved by specially designed automatic cross-connection equipment, which uses the principle of time-slot interchange employed by digital telephone-network switching centres. Cross-connection is controlled remotely from a network control centre.

By using a relatively small number of separate cross-connection centres, this approach allows digital leased private circuit provision to be independent of the rate of introduction of the more complex and costly digital telephone switching centres, and does not require additional equipment to be provided at these centres against a possible private circuit need. In addition, circuit capacity for leased private circuits is concentrated on fewer routes to make the most efficient use of transmission systems and lessen the effects of forecasting error.

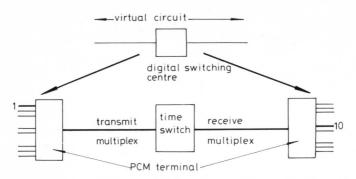

Fig. 5.2   *Semi-permanently switched private-circuit connection. Information transmitted in time slot 1 is delayed in the time switch by 9 time-slot periods so that it is received in time-slot 10 for a predetermined period instead of on a per-call basis.*

The public-network transmission-system multiplexing hierarchy allows 'circuits' of greater frequency bandwidth (analogue) or transmission bit rate (digital) to be extracted from transmission systems, to provide a leased private connection offering users more capacity than a single telephone circuit. The user is offered a wide frequency or bit-rate band which may be utilised in total for one purpose, for example a video circuit, or multiplexed as required at the user's premises from separate speech and/or data channels; the latter can be tailored to any desired transmission rate (subject to the availability of suitable multiplexing equipment) within the overall capacity provided. Digital circuits are commonly provided to offer a user the equivalent capacity of, for example, a public-network PCM system, namely 1.5 or 2.0 Mbit/s. Existing local-network cables may not provide a suitable transmission environment for the local ends of such circuits, however, necessitating the provision of new cable or an alternative means of transmission to link the user's premises to the main transmission network (see Section 4.1).

5.2 Private branch exchanges (PBXs) and private telephone networks

The term 'private *automatic* branch exchange' (PABX) was introduced in the United Kingdom and some other countries to distinguish PBXs using automatic switching from the original *manually* operated PBXs (PMBXs). With the virtual disappearance of the latter from the market, the letter A in PABX has tended to be dropped by many manufacturers and users. In line with this trend, all references to PBXs in this book imply the automatically switched variety, prefixes being used to distinguish between, for example, analogue and digital PBXs. Most new PBXs are now of the digital type, sometimes referred to as *third generation*, and both these and the modern analogue *second-generation* PBXs employ microelectronic circuitry and stored-program computer control (SPC) as in the public-network switching field. (The term *first generation* is applied to the obsolescent electromechanical analogue PBXs.) It is useful, therefore, to examine the modular structure and working of typical PBXs of the current second and third generations, designed – whether digital or analogue in operation – for connection to the public network over analogue local circuits. (*Fourth-generation* digital PBXs, now emerging and referred to in Section 5.4, interface with the ISDN over digital local circuits.)

There are differences (apart from capacity) between large and small PBXs – for example, in terms of the range of facilities offered – but all, whether analogue or digital in operation, can be broken down into a number of functional elements, as shown typically in Fig. 5.3. These are (a) a control unit, including the computer which controls call set-up and clear-down, (b) a switch unit, (c) one or more extension interface units, (d) an operator's console, and (e) various miscellaneous items which may collectively be termed the peripherals. The purpose of these elements is explained in the following paragraphs.

The *control unit* is most likely to be built around a purpose-designed computer – usually referred to as a *processor* – rather than an adapted general-purpose model, using commonly available microcomputer 'chips' at its heart. In the larger PBXs at least, control is fully duplicated for added security; a common approach is to operate two identical processors in parallel. Only one actually assumes control at any particular time and the other is ready to take over, without interruption, should the controlling processor fail (the so-called 'hot standby' approach). The computer program that runs on the processor(s) and actually exercises control contains all information appropriate to the operation of the PBX and its extensions. This control program is a separate entity which enables the manufacturer to introduce program updates to correct unforeseen problems, introduce new facilities, etc., and also permits the easy changing of facilities available to particular extensions. Most control programs, especially those used in the larger PBXs, are to some extent designed to make the PBX 'self-healing', in that a limited range of minor faults can be tolerated without affecting the capability of the PBX to handle calls. The control program, or in some cases a separate optional program, also records information on traffic levels, the number of calls handled, fault conditions, etc. The amount and detail of the information provided varies considerably; it is likely to be more comprehensive for a large PBX, where printed reports may be produced on a telex-type terminal.

The *switch unit* includes several stages of switching to link extensions, exchange lines and other peripherals as required. All switching stages in an analogue PBX will be space switched, but a digital PBX must include at least one stage of time switching, using the public-network switching-centre principles explained in Section 2.8. A small PBX will normally be equipped with a switch unit of fixed capacity, able to meet the requirements of the maximum number of extensions catered for by the PBX design, whereas a large PBX may be equipped with modular switch units, permitting capacity to be added in fixed amounts up to the maximum design size.

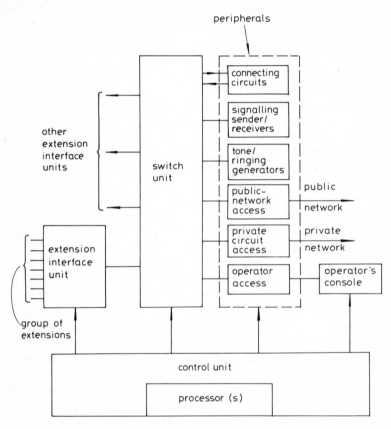

Fig. 5.3 *Major elements of a typical PBX*

An *extension interface unit* connects a group of extensions to the PBX, each of its line circuits performing the signalling and ringing/tone distribution functions for one extension. Multifrequency signalling is often employed, as in public networks, to improve call set-up time. In a digital PBX, the analogue-to-digital conversion may take place either (a) at the interface unit or (b) at the extension telephone (the short **PBX** line lengths do not normally require the special digital transmission techniques necessary for public-network local circuits, referred to in Section 4.1); a **PBX** may be designed to provide for either or both possibilities. Digital multiplexing is accomplished at the interface unit. In some cases an extension interface unit may be separately controlled by its own microprocessor.

The *operator's console* is provided on the basis of one per operator; one will suffice for a small **PBX**, whereas a number will be needed to cope with the demand imposed by a very large **PBX**. Very small PBXs, known as *key systems*, do not require an operator's console; instead, extension telephones are equipped with additional keys to obtain direct connection to the public network and other extensions.

Finally, the *peripherals* include circuits for terminating public-network lines, external private circuits (sometimes referred to in the **PBX** context as *tie lines*) and the operators console(s). Other essential peripherals include circuits needed to interconnect two extensions and circuits used only during call set-up, such as signalling sender/receivers and tone/ringing generators; these are normally concentrated in a central pool, being connected to lines via the switch unit when instructed by the control program.

All the functional elements with the exception of the operator's console are housed together; a relatively small table-height or wall-mounted cabinet suffices for a **PBX** offering tens of extensions, whereas a number of cabinets 2 m high are required for a **PBX** offering many thousands of extensions. Individual **PBX** designs differ; there may be a different functional split, and some elements may be supplied by the manufacturer as a complete invariable entity while others may be built up from standard modules, supplied according to the number of extensions and public-network lines required and/or traffic and call-handling demand. Modularity permits a **PBX** to be tailored more precisely, and thus more economically, to a user's requirements, enabling forecast growth in demand to be met by the addition of extra modules as required. In general, as might be expected, larger PBXs offer greater modularity.

Modern PBXs offer a range of facilities which are not yet available to individual users of the public telephone network on a universal basis. These facilities include the automatic selection of least-cost routings where alternative public-network outlets are available; personal directory number storage combined with short-code keying; extensive call-transfer options; a ring-back-when-free feature; and so on.

A **PBX** of the type described in the preceding paragraphs is primarily designed to handle telephone traffic. It operates on a circuit-switched basis, in common with public telephone network switching centres, providing 3 kHz (analogue) or 64 kbit/s (digital) switched paths. Nevertheless, as in the public network (Section 3.1), non-telephone (sometimes referred to as non-voice) traffic can be transmitted over such paths. Computer data and text information can, therefore, be exchanged between extensions, or between an extension and the public net-

work(s), either directly in the case of digital PBXs or via modems in the case of
analogue PBXs. Data transmission rates are limited by the extension line length
and type of transmission involved, as in the public networks, and a digital PBX
equipped for digital operation from extension terminal to extension terminal
offers the advantage of data transfer at higher bit rates than would be possible
over analogue extension lines.

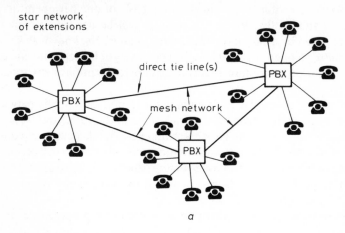

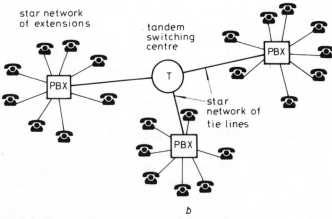

Fig. 5.4 *Private telephone network:*
(a) Typical star/mesh configuration
(b) Tandem-switched network with star/star configuration

A number of PBXs can be linked by tie lines carried on individual leased
circuits or circuits multiplexed over leased bandwidth on a public network, by
private microwave radio links, etc. to form a private telephone network. As in
a public telephone network, the extensions served by each PBX are connected

to it in a *star* configuration but the inter-PBX tie lines are effectively configured as a *mesh* network (Fig. 5.4a). PBXs generally, however, are equipped only to switch traffic between their own extensions and either the public network or tie lines; they are not able to act as an intermediate or *tandem switching* point between two other PBXs. For this to be possible an additional public-network-type tandem switching centre is necessary (Fig. 5.4b), as discussed in Section 8.1.

The private circuits or wideband capacity used for interconnection may cater for either analogue or digital transmission. The latter offers improved, noise-free transmission quality even where, as initially in the established public networks, analogue-type space switching is employed. The combination of digital transmission and digital switching reduces transmission loss and offers scope for transferring data at higher speed throughout the private network, subject only to a 64 kbit/s upper limit and to the restriction imposed by call set-up time, common to any circuit-switched network. Signalling over long-distance private circuits requires the use of techniques similar to those employed in the public networks (Sections 2.2 and 2.4). Common-channel signalling closely allied to the public-network CCITT no. 7 (Section 2.8) is used between digital SPC PBXs; the system is known in the UK as the *digital private-network signalling system* (DPNSS).

A description of PBXs would not be complete without a reference to *centrex* – the provision of modern PBX facilities from a public-network switching centre without the need for on-site PBX equipment. This service, available for many years in the USA and now being offered in the UK, can provide a cost-effective alternative to PBX ownership or leasing for the small business.

5.3 Private data networks and the local area network (LAN)

A private data network is optimised solely for the rapid transfer of data traffic, as opposed to a basically telephone network which is subject to call set-up delay and can only carry data traffic at limited transmission rates end-to-end. A private telephone network as described in the previous section is normally structured on a mixed star and mesh basis, every node within the mesh being a switching point and PBXs being the hub of a star network of extensions, so that every extension is able to communicate with all other extensions. Data communication does not necessarily require that every extension on the network, normally referred to in data-network terminology as a *station*, should be able to communicate with all other stations; thus a switching facility is not always necessary. Large message- or packet-switched private data networks equipped with public-network-type switching centres do exist, but in most cases a more simple network structure is adequate.

The simplest data network consists of one point-to-point link between, for example, a data terminal and a computer (Fig. 5.5a). Where a number of separate data terminals need to communicate with a central computer but not

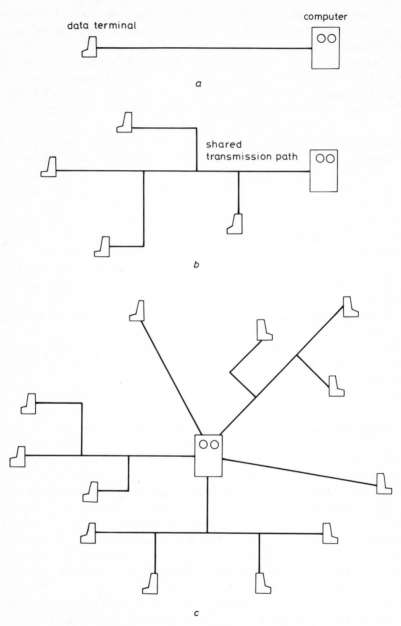

Fig. 5.5 *Simple private data network configurations:*
(a) Point-to-point
(b) Multidrop: terminals can communicate with computer but not with each other.
Computer polls terminals in rotation to exchange data with each in turn
(c) Star: combined multidrops and point-to-point. Terminals can still only com-
muniate with computer, not with each other

with each other, a *multidrop* network is appropriate (Fig. 5.5b). There is only one shared transmission path in a multidrop network; the computer may transmit over this to all terminals simultaneously, but the terminals can only transmit to the computer one at a time. The computer uses a special protocol to contact the terminals in a predetermined order and receive data from each in turn; this is known as a *polling mechanism*. A number of point-to-point and/or multidrop subnetworks can be combined into a *star* network (Fig. 5.5c), each accessing the computer via a separate connection point known as a *port*. In large networks, communications control may be devolved to one or more separate interface computers known as *front-end processors*, freeing more central-computer processing power for the main computing tasks.

A star network of this type may be adequate in a straightforward single-computer multiterminal situation; however, for more complex data networks which require the sharing of common resources, such as one or more mainframe computers and their data bases, printers, etc., it has two major limitations. First, a switching facility would be needed at the hub to enable all stations to communicate directly with each other. Secondly, the network is very vulnerable to system or link failure because no alternative communication paths are available. A *mesh* network, as used to interconnect switching centres in public networks, could overcome both these problems. All stations in a mesh network are equal in status and could, if fully connected, communicate with each other by accessing the appropriate link; alternative paths could also be set up between any two stations should a link fail. However, a mesh-type data network can be very expensive to establish and operate unless the traffic volume between stations is consistently high and evenly distributed, because of the large number of links needed; transmission over a public packet-switched data network may well offer a more economic and efficient alternative.

Two unique network configurations have been developed to permit limited-distance direct communication between *any* two stations without the need for a switching centre. Both employ (a) a common transmission path and (b) a form of distributed packet switching to route messages to the required station. The transmission path is cabled throughout the area to be served so that additional stations (up to a limit of, typically, around 100 stations per network) can be 'plugged in' to it wherever the need arises. Data networks of this type are known by the generic term *local area network* (LAN) since they are designed specifically for a single site or several nearby sites, where the distances over which data needs to be transmitted are normally quite short and rarely more than a few kilometres. LANs provide for high rates of error-free data transfer and are ideally suited to a general-purpose data network which includes shared-access computers, printers, etc.

One of the possible LAN configurations is the *ring* or *loop* network, where the stations are linked to *repeaters* spaced around a single closed loop, either directly or by means of branching connections (Fig. 5.6a). Transmission can be (a) in one direction round the ring only, in which case there is no alternative path for the transfer of data should a link fail, or (b) in either direction, when there is just one alternative path – the opposite direction – if a link fails. A duplicate standby ring can be provided as a safety measure, although this could well prove to be equally vulnerable to failure if it follows the same physical path as the main loop. An alternative safeguard against link failure can be provided by configuring the network as a *star-connected ring* (Fig. 5.6b); this allows a failed link to

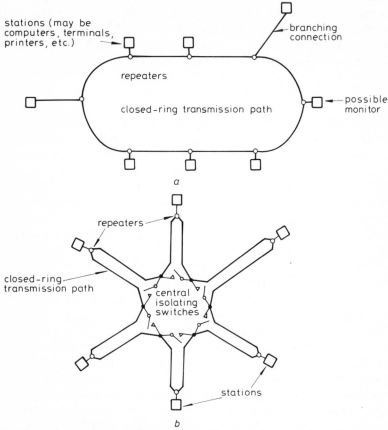

Fig. 5.6 *Ring or loop LAN:*
(a) Normal configuration
(b) Star-connected ring. The ring is physically constrained to a small central area between repeaters. In the event of cable breakage or repeater failure, each spur can be isolated by a central switch to maintain the loop for other stations

be easily bypassed. A complex set of protocols performs the control and distributed switching functions to ensure that data signals can be transferred round the common transmission path between specific stations as required. A monitor station can be connected to the ring to check for correct operation and raise an alarm if any irregularity is detected.

The other possible LAN configuration is the *bus* network, which utilises a single linear transmission path known as a bus. Several such paths may be linked by branching repeaters and each station is connected to a transmission path via an interfacing device known as a *transceiver* (Fig. 5.7). A bus network with branching connections is sometimes referred to as a *tree* network, although strictly speaking a true tree network has a controlling/switching hub at the 'trunk' end (see broadband LANs later in this section). Again, a complex set of protocols performs the control and distributed switching functions to enable all stations to communicate with each other. A single bus cannot offer an alternative path in the event of link failure, so a duplicate standby bus must be provided if continuity of service is considered to be of importance.

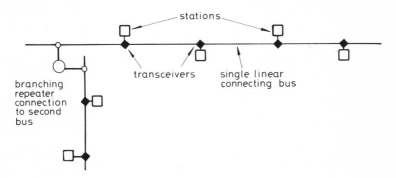

Fig. 5.7 *Bus LAN*

Numerous different, mostly proprietary, sets of protocols have been developed for use with LANs. Each set of protocols is designed for a particular network configuration – either the ring or the bus – and can be further classified according to the methods used (a) to control access to the common transmission path and the transfer of data over it, and (b) to provide and utilise this transmission path.

Three methods of controlling data transfer, two of them very similar in principle, have so far been evolved. All rely on the transmission of data as addressed packets with error checking and retransmission incorporated, as in the public packet-switched network (Section 3.3), but packet length and makeup vary considerably between protocols.

The first, a unique method known as *carrier-sense multiple access* (CSMA), can be applied to ring or bus networks. Stations that wish to transmit have to contend for use of the transmission path on a first-come, first-served basis, each station listening in to the transmission path (sensing the carrier) to check that it is free before commencing to transmit (Fig. 5.8). Once transmitted, the packet address ensures that thé data is accepted by the required station and no other. The situation can arise where two stations seize the transmission path simultaneously and transmit at the same time; the two clashing data packets would then be inseparable from each other and thus indecipherable. To overcome this problem, an additional feature known as *collision detect* can be incorporated; this gives rise to the abbreviation CSMA/CD. The collision-detect feature causes the simultaneous transmissions to be halted and a jamming signal to be sent over the transmission path, to stop further attempts to transmit until a short interval after the path is clear of the garbled data. Each station involved

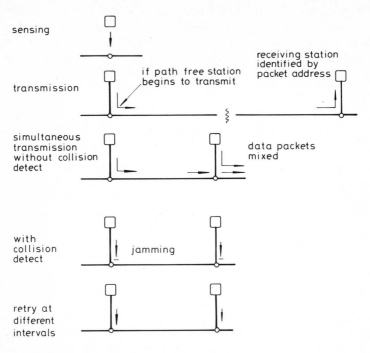

Fig. 5.8 *Carrier-sense multiple-access data-transfer technique*

in the clash then automatically tries again after a random, and thus different, interval. The stations are allowed to retry until data transfer is successfully completed or the transfer is automatically rejected as unsuccessful after a fixed maximum number of attempts. Rejection, which is only likely to be encountered under conditions of extreme network congestion, necessitates a completely fresh attempt by the unsuccessful station. The original and most well established proprietary system using CSMA/CD is Ethernet, which uses a bus-type network and was developed by the US Xerox Corporation in the early 1970s.

One of the other two similar data-transfer methods is suited only to ring networks and uses the *empty-slot* technique. With the empty-slot technique, developed at Cambridge University in the UK and typified by the well known Cambridge ring system, spaces for data packets (in other words empty packets, which are in fact empty time slots) permanently circulate the ring whilst the system is switched on. A station wishing to transmit seizes the first passing empty packet, marks it as full and fills it with data. The receiving station, as identified by the packet address, copies and thus receives the data; the full packet carries on round the ring until it returns to the originating station, where it is stripped of data and remarked as empty (Fig. 5.9). The empty packet then carries on to the next station, from which point onwards it can again be seized for a new data transfer. In this way, a station is (a) notified that its data packet has been received and (b) prevented from immediately reseizing the same time slot each time round to hog the network for its own transmission. Since each data packet is transmitted separately, a complete data message is transferred over a number of separate transmissions. Other safeguards are also incorporated in the protocols; for example, to ensure that, in the event of the intended receiving station being engaged, a full packet cannot circulate the ring indefinitely.

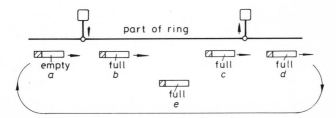

Fig. 5.9 *Empty-slot data-transfer technique:*
(a) Succession of empty packets continuously circulates ring
(b) Transmitting station fills empty packet
(c) Receiving station identified by packet address
(d) Receiving station reads full packet but does not empty it
(e) Full packet returns to transmitting station to be emptied

The third data-transfer method, which is similar to the empty-slot system, is suited to ring or bus networks and uses the *token-access* or *token-passing* technique. With this method a special data packet, comprising a unique bit pattern known as a *token* which can never occur in normal data transfers, is present on the ring or bus whenever the network is free. A station wishing to transmit removes this token from the transmission path to indicate to other stations that a transmission is in progress, and restores the token at the end of its transmission (Fig. 5.10). The amount of data that can be transferred by one station during any one transmission is automatically limited, usually to a single packet, to prevent hogging of the network. As with the empty-slot method, a number of separate transmissions are necessary to transfer a complete data message. The US International Business Machines Corporation (IBM) is a forerunner in the field of token-based technology.

LANs of the types so far described are collectively known as *baseband* systems. They operate at data transmission speeds ranging from several hundred kbit/s to 10–12 Mbit/s, although most systems are able to accept traffic at only a small fraction of these bit rates. High operating speeds necessitate repeaters at intervals along a lengthy transmission path, as already described for public-network digital transmission systems (Section 2.4). Systems operating at the lower speeds utilise ordinary metallic-pair cable for the tranmission path, whereas higher-speed systems may require coaxial cable. Although, theoretically, stations can be connected at any point on the ring or bus cable, in practice most systems require the actual positioning to be at precise equidistant points. It is also possible to obtain LAN systems utilising light transmission over optical fibre.

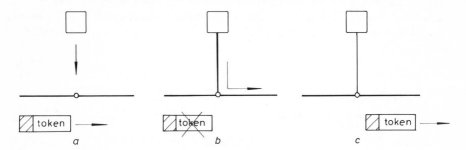

Fig. 5.10   *Token-access or token-passing data-transfer technique:*
(a) Token present on ring or bus whenever transmission path is free
(b) Station wishing to transmit checks for presence of token. If present, removes token and transmits data
(c) Station restores token at end of transmission

An entirely different class of LAN has been developed from ideas first used for cable-television distribution; this has permitted advantage to be taken of mass-produced cable-television components, particularly in the USA. These LANs are collectively known as *broadband* systems to distinguish them from baseband LANs. The broadband LAN concept incorporates the techniques of accessing a common transmission path, as used in baseband systems, with telephone-network multiplexing principles. The available bandwidth on the transmission path is split into a number of separate, independent channels by means of frequency-division multiplexing (Section 2.2). The channels can be used for (a) different and quite separate services, for example speech or video as well as data, (b) data transmission employing different access techniques, for example CSMA or token access (although the principles used for baseband systems have to be adapted to the broadband concept), and (c) dedicated private point-to-point circuits.

The cable-television ancestry of broadband systems is evident in the use of a tree network (several of which may be combined into a star-type network) and in the terminology employed. Transmission occurs in both directions, termed 'upstream' and 'downstream'; a message is first transmitted upstream by the originating station to the head end and then repeated downstream to the required destination (Fig. 5.11). The bidirectionality is achieved by using either (a) two separate cables, one for each direction, or (b) a single cable over which each direction is allocated a separate frequency band. In either case transmission in each direction is split over several channels. A device known as a *transponder* is located at the head end, its function being to receive all signals transmitted on the upstream channels and retransmit them on the appropriate downstream channels; in a single-cable system this necessitates the translation of the signals to a different frequency band. Since all signals must pass through the transponder, it is critical to the operation of a broadband LAN and is, therefore, usually duplicated against the possibility of failure. Overall control of the system is vested in a separate host computer, which is connected to the network in the same way as any other station.

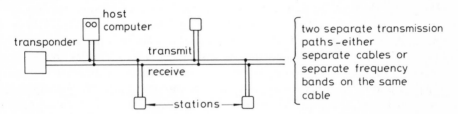

Fig. 5.11 *Broadband LAN. Transmit and receive transmission paths are each divided into a number of separate frequency bands (channels) which can be used for different services*

Broadband LANs normally employ coaxial or optical-fibre cable and are able to exploit the potential frequency bandwidth of the cable to a much greater extent than baseband systems, operating at very high data-transfer rates to offer improved throughput capability over longer distances (50 km links are technically feasible). However, changes to a broadband network are more difficult to implement once it has been installed and, in addition, broadband systems are more complex than baseband systems. Although this complexity is to some extent offset by the availability of cheap components, broadband systems are usually more expensive to establish and maintain than baseband systems of equivalent capacity. Broadband systems are, therefore, best suited to high-volume data applications where short response times are critical and where separate voice and video services are also required (see also Section 5.4).

For both baseband and broadband LANs, the maximum number of stations and repeatered branching connections is determined by the overall transmission limitations of each particular system. Since a repeater merely regenerates received bits, these limitations apply to the complete network including any branching connections, which restricts the length of each branching connection to no more than a few kilometres. Wider geographical coverage can only be achieved by linking together a number of LANs, the complete data network then being known as a *wide area network* (WAN). This networking concept effectively removes the restrictions on network area and number of stations which apply to an individual LAN.

WAN inter-LAN links can be operated on a broadcast or point-to-point basis. With *broadcast* working, data frames are transmitted throughout the whole linking network, being selectively received according to their address; this makes for technical simplicity but increased link capacity is required. With *point-to-point* working, an additional interface is required between a LAN and its link to other LANs. Two different types of interface equipment are available, the *router* and the *data-link bridge* (often simplified to 'bridge'), but they cannot be mixed: a network must be interconnected using either wholly routers or wholly bridges. Both store data transmitted over the LAN to which they are connected and forward it over the link, but they differ in the way they function. The router depends on local LAN stations to address to it only data frames intended for distant LANs, whereas the bridge interrogates *all* data frames transmitted over the local LAN, filtering out those destined for local stations and forwarding only those intended for distant LANs. In a bridge-connected network of LANs, local stations simply address data frames to all other stations (including those on other LANs) as if they are on the same LAN; this requires additional intelligence in the bridges. In a router-connected network, data frames intended for a station on a distant LAN are addressed differently; this requires additional intelligence in the stations. There is no difference in the way routers and bridges receive data frames over the linking circuit from distant LANs; in both cases only data frames destined for stations on the associated local LAN are involved. Essentially, in a bridge-connected network the bridges keep the traffic local to one LAN from interfering with the traffic local to another; in a router-connected network this responsibility falls on the local stations which are, therefore, required to use more complex protocols. Since router-to-router or bridge-to-bridge links are required to carry less traffic than each of the LANs to which they are connected, the transmission rate used over a link can be slower than the transmission rate of either of the LANs to which it is connected.

A common factor with all the private data networks described, whether LAN based or unswitched, is that they are essentially computer networks. The most important network traffic parameters, therefore, relate to *message pairs* to and from the computer and the average time in which a response can be expected (see Section 6.5.2). Traffic flow is directly measurable in terms of the number of bits transferred per second (bit/s) and conversion to telecommunications-based traffic-flow units would be pointless.

Data-network design is complicated by the number of proprietary protocol and interface standards which exist, in addition to the internationally agreed standards designed to lead to open systems interconnection (OSI). A wide range of equipment is available to perform protocol conversion, but the very diversity of equipment options available makes network design a province for the data-communications expert. The most straightforward solution, which even an independent consultant may recommend, is often a complete proprietary network provided by one manufacturer. However, a single-manufacturer network may not provide the most economic or technically advanced long-term solution and can lead to dependence upon that manufacturer for years to come, since future updating will most easily be achieved with compatible equipment from the same source. Both consultant and supplier(s) must, therefore, be chosen with care if an optimum network solution is to be obtained: the former for a proven record of independence and technical competence and the latter, especially if one only is involved, for financial stability as well as technical and economic product superiority.

5.4 Integrated speech and data on PBXs and LANs

Section 4.1 explained how public telecommunications networks are progressing towards the integrated-services digital network (ISDN) concept, where users will have a single network-access point from which to obtain all telecommunications services. It is already theoretically feasible to carry all these services – speech, data, facsimile and video – on a single integrated network, but any change from the established separate networks must await the technological advance from theory to production systems. Even then, the change can only take place over a lengthy period because of the huge amount of capital investment involved in once again replacing existing equipment worldwide. Similar pressures for integration are at work in the business-dominated world of private networks, but here the constraints are fewer and the capital investment requirements are spread more widely. It is to be expected, therefore, that service and network integration will be achievable in private networks long before similar changes to public networks can be implemented on a nationwide basis. Moreover, the OSI objective of full interconnectivity for data communications, although still a long way from being met, will be brought that much closer by the updating of existing private systems and networks to new, common standards. Integration, for private as well as public networks, has as its primary objective the integration of speech and data communications; since facsimile and some video information can be transported over speech circuits, these forms of communication are included by default. A secondary objective is the addition of high-quality video communication to these services.

Although present-day PBXs operate on a circuit-switched basis and are designed primarily for telephone traffic, they can carry data and other non-telephone traffic provided it is suited to telephone-call-type handling with the inevitable call set-up delay. There are a number of third-generation digital PBXs on the market for which some degree of intra-PBX speech and data integration is claimed, but these claims need to be examined carefully, especially from the point of view of their limitations and the cost of the necessary ancillary equipment. Only a fourth-generation PBX, known as an *integrated-services PBX* (ISPBX), provides both speech/data integration over common extension circuits and public ISDN compatibility. The role of the ISPBX has been extended beyond the basic switching and public-network interface functions of a PBX; it acts as a specialised gateway to a wide range of telecommunications services, providing such additional data-communications functions as store-and-forward data transfer, protocol conversion, etc. Extension/extension and extension/public-network digital transmission paths are provided, access to the public network being obtained via a PCM link. Common-channel signalling is used between private and public networks to permit a much greater range of facility information to be exchanged, using a system based on the principle of the public-network CCITT no. 7; in the UK a system closely allied to DPNSS (Section 5.2), known as the *digital access signalling system* (DASS), has been developed for this purpose.

PBX operation on a circuit-switched basis remains a barrier to the integration of *all* data traffic with speech, since it is incompatible with (a) heavy data-traffic flow for long periods, which would swamp the PBX traffic-carrying capacity, and (b) data traffic requiring a near-instantaneous response time. In addition, the fixed 64 kbit/s transmission paths impose an upper limit on data transmission rates and are unsuited to the transmission of high-quality video signals.

By contrast, LANs operate on a packet-switched basis and are expressly designed for the efficient handling of all types of data traffic. Although broadband LANs can also carry speech traffic, this is on completely separate channels and in no way offers the possibility of integrated speech and data over a common path. The packetising of speech presents no technical difficulty but its transmission must not be delayed by more than a few milliseconds, requiring priority over all other traffic; this is a feature not offered by the LAN protocols used in currently available systems.

The limitations of circuit switching preclude the possibility of its use for the transmission of all data traffic. The problem of transmitting speech over a packet-switching system is not insoluble, however; more advanced LAN protocols have already been shown to offer a theoretical solution and production systems will quickly follow once a demand has been established. This does not imply the demise of the PBX as a business telecommunications system, however; PBXs can equally well employ a form of packet switching, using similar protocols to integrated-services LANs, and *fifth-generation* PBXs based on this concept are already under development. The long-term solution for both LANs and PBXs would appear to lie with the enhanced form of packet switching known as *variable-bit-rate switching*, as in public networks (Section 4.2). Variable-bit-rate switching offers scope for the integrated transmission and switching of high-quality video signals as well as speech, data and facsimile. In addition, by dynamically allocating to each switched transmission path only sufficient bandwidth for the bit rate required by that particular transaction, the system is able to use the available capacity with maximum efficiency.

Given that both LANs and PBXs will be able to offer comparable standards of service integration, the choice between them will have to be made on factors which result from their differing network concepts; the LAN with its distributed switching and ring or bus network, and the PBX with its centralised switching and star network. Technically there would appear to be little to choose between the two systems, and it is not yet possible to predict which might offer the most economic solution. An installed cabling base might well prove to be the deciding factor by determining the network structure and thus the system choice.

Chapter 6

Traffic measurement, forecasting and equipment quantities

6.1 The purpose of traffic measurement

In Chapter 1 it was established that any telecommunications system, from a single circuit to an entire network, has a finite capacity for carrying traffic. As the capacity limit is approached, the way in which users of a switched network are affected will depend on the type of network involved, as summarised in the following list:

(a) In a *circuit-switched telephone network* or similar telex network, free circuits will not be available to new call attempts which will fail in growing numbers. Unsuccessful callers will make repeat attempts to establish a connection, adding to the number of call attempts being made. The result will be increased network congestion, with more and more callers unable to establish the desired connections.

(b) In a *message-switched data network* such as a modern telex network, the addition of a store-and-forward facility enables circuits to be more efficiently loaded and operated more closely to the capacity limit. Initially, a lack of free circuits will lead to an increasing number of messages being stored at a switching centre awaiting transmission, and hence delayed in transfer across the network, but messages will continue to be accepted. As the network storage fills up, delays will increase until there is no longer sufficient storage space for new messages; users are then faced with even more delay because they cannot resume transmission until storage space at their local switching centre has been freed by the transmission of a message from the store.

(c) In a *packet-switched data network*, users will also experience increasing network delays, as in a message-switched network but affecting the completion of messages in transit rather than necessarily their whole.

In order to avoid these problems without incurring the economic penalties of over provision, equipment quantities throughout a switched telecommunications network must be closely matched to the actual traffic demand. For a new network, a good match is initially difficult to achieve because it depends solely upon a forecast which, however well researched, seldom reflects the real demand with sufficient accuracy. Once a network is operational, however, traffic measurements permit the real demand to be verified and contribute to the successful management of the network in two ways. In the short term, they enable unforeseen localised traffic growth to be detected so that, if and when necessary, additional network capacity may be brought into use. They also allow trends of traffic growth (or decline) to be established and projected into the future; a forecast of longer-term traffic demand can then be produced to make it possible to plan the future development of the network on a sound economic basis.

6.2 Telephone-network traffic measurement and forecasting

The idea of measuring traffic in terms of traffic-flow units was first conceived with the sole objective of ensuring the cost-efficient handling of telephone calls on a circuit-switched network. Other types of telecommunications traffic are now carried on the same circuit-switched network in addition to telephone calls but, since all traffic is in the form of separate successive *calls*, traffic measurement on the basis of erlang or CCS units remains a valid concept.

If the holding time of every individual call had to be measured, this would have made traffic measurement in the pre-electronic era unnecessarily complex and expensive. In practice, therefore, a simplified approach was adopted by network operators, to provide *average traffic-flow* figures which could be interpreted with sufficient accuracy for their needs. On this basis, a representative sample of call holding times, covering both successful and unsuccessful calls, is used to calculate an *average call holding time*. It is then only necessary to count, over a period of one hour, the number of times each circuit on a particular traffic route is engaged, and multiply this number by the average call holding time to give the average traffic flow on the route in erlangs or CCS. Private networks pose a more difficult problem than public networks, particularly for within-network traffic, since the sample must be taken from a much smaller population and is, therefore, statistically less reliable. Because private networks are mostly relatively small in comparison with public networks, calls which last longer than the average call holding time can have a more significant effect on network performance, although this is counterbalanced to some extent by the fact that the reserve capacity provided by each circuit represents a much larger proportion of the total network capacity.

Traffic measurement is carried out at the switching centres, the measuring and recording equipment forming an integral part of each centre. The traffic on each traffic route *outgoing* from a switching centre is measured separately, the engaged periods on all circuits comprising a route being counted and added to give a route total (traffic *incoming* to the switching centre is measured at the distant end on the traffic routes concerned). Internal measurements are also taken within each switching centre. Measurements are not required on a continuous basis, being necessary only for the daily *busy hour* of the routes etc. under observation. This is preselected on the basis of past measurements and the recording and processing of measured traffic flows in modern electronic switching centres is fully automatised. In addition to engaged circuits, a count of call attempts (usually referred to as *busy-hour call attempts* (BHCA)) is also important since these exceed the number of successful calls; it is attempts rather than actual calls which determine the loading on the call-control processor and its associated equipment. Additional facilities, such as (a) sample measurements of call duration and destination and (b) records of times when congestion is encountered, will vary between different manufacturers and may be tailored to suit the requirements of a particular network management procedure. It must be emphasised that average traffic flow measurements cannot indicate a precise instantaneous value of traffic; hence additional information in the form of congestion records is essential if a complete picture of traffic bottlenecks is to be obtained.

The *periodicity* of traffic measurements must be such as to provide statistically reliable results. There are differing views among public network operators as to the optimum periodicity, but generally measurements are (a) confined to working days, excluding Saturdays and Sundays, and (b) taken over a number of successive working days (typically from 5 to 20) at intervals throughout the year (typically 12 times a year for the shorter periods and up to 8 times a year for the longer periods).

Traffic *forecasting* is tailored to the network concerned and is also influenced in detail by the amount of information available to the forecaster; in essence, the approach used is likely to be determined by whether the network is large or small, public or private. However, the principles involved in preparing a forecast for a *traffic route* in an established *public telephone network* provide a useful introduction to the subject. The essential prerequisites for such a forecast are estimates of (a) the timing of the daily busy hour for that route, based upon past measurements, and (b) the average call holding time, derived from network-wide sample measurements during this predicted busy hour. The number of engaged circuits during the route busy hour is then counted on several successive working days, which together constitute a measurement period. These figures enable the average busy-hour traffic flow over the measurement period to be calculated in erlangs or CCS. An overloaded route is unable to carry all the traffic offered to it; hence the measured traffic level under congestion conditions may not accurately reflect the true traffic-flow potential; it is, therefore, important to take account of the presence of congestion on a route when interpreting measured figures. The average traffic flow quantities from successive measurement periods throughout the year are then plotted on a graph to reveal the likely trend of future traffic flow demand. This trend line, being based solely on historical evidence, must finally be adjusted to take account of the likely effect of any planned or probable future events that could influence the forecast. Such events may have an influence which is (a) network-wide, for example a planned tariff increase, or (b) purely local, for example a proposed new industrial development in an area served by the traffic route.

A *switching centre* is a composite entity, required to perform a variety of functions for which the demand does not grow uniformly; different types of equipment module, therefore, need to be augmented by different amounts, depending on their particular function and location within the centre. Because of this multi-dimensional requirement, traffic forecasts for a public-network switching centre need to be much more comprehensive than those for an individual traffic route. A forecast must include (a) the busy-hour call-attempt (BHCA) rate, essential to ensure the adequacy of the call-control processor and allied equipment to handle all call attempts presented to it; (b) the traffic on all routes connected to the switching centre; and (c), where a local switching centre is concerned, the number and type of users to be served.

Forecasting is normally undertaken on an annual basis, but how far forward the forecasts extend will depend on the purpose for which they are intended. One year ahead normally suffices for the augmentation of circuits, since transmission systems are installed to provide capacity in excess of the immediate need, as indicated by a longer-term forecast, and it is a relatively speedy process to connect additional circuits within that capacity. Longer-term forecasts are required for (a) the transmission network and (b) switching centres, to allow time for any additional equipment to be ordered and installed. Forecasts for the provision of equipment accommodation must look many years into the future, since building work and even site purchase may need to be planned and executed.

An established *private telephone network* will not be affected by all these considerations, but the principles remain valid. The forecasting technique described is heavily dependent upon past measurements; this must be so because the public-network forecaster has no idea of the demands each individual user of the network will make upon it. The private-network forecaster is much better placed in this respect, being concerned only with the network serving a single organisation. Although any traffic measurements available may not be as comprehensive or reliable as for a public network, the private-network forecaster has access to information not only on all the terminal equipment and its capability but also on the jobs of the users, and is thus in a position to assess future traffic levels with greater confidence. It is not essential to provide a forecast in terms of traffic-flow units for a small private network; a simple forecast of the expected number of calls and average call duration will usually suffice.

Where a *new* network is to be established, a forecast of traffic demand often has to be constructed without the benefit of historical evidence (although information derived from existing networks of a similar type and size may be available in some circumstances). Clearly, this is a much more difficult task which is inevitably more prone to error. The starting point for such a forecast, whether it is for a public or private network, must be a detailed analysis of the market to be served. Before the format of the forecast can be decided upon, however, the forecaster must be aware of the exact purpose for which the forecast is intended. If a particular network configuration is planned – for example, the simple case of a single PBX – the forecast can be precisely tailored to this need, as it would be for the augmentation of an established network. If, however, a more complex interconnection pattern is required, a much more general forecast is needed so that alternative network configurations can be considered. Once such a network is operational, traffic measurements are quickly available to enable the initial forecasts to be adjusted to reflect actual demand.

New private telephone networks are increasingly likely to be required to handle a mixture of telephone, facsimile and data traffic. A multipurpose network of this type may be needed for an entirely new enterprise or it may be intended to replace an existing, outdated network. The possible network configuration may be restricted, for example, because of the existence of usable cabling, or the requirement may be open ended to enable an optimum network to be designed. Although measurements from established networks similar to that proposed may not be readily available, and those from an outdated network may be of little use when forecasting for a new network, the private-network forecaster can readily collect estimates of future needs from within the single user organisation concerned. The collected information must be separated into calls (a) to destinations within the private network, with details of frequently required connections, and (b) to and from external public-network destinations. Where it will be necessary to compare the cost of different network configurations and routing options to determine the cheapest solution, the work will be facilitated if within-network traffic is aggregated in terms of calls between

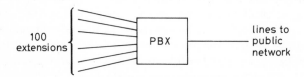

Fig. 6.1   *Traffic flow in an example PBX (See text)*

areas of common interest, such as office departments and/or buildings, rather than individual users. The period(s) when traffic is likely to be at its heaviest can then be identified, allowing a choice to be made between a single network-wide busy hour or different busy hours for parts of the network. It is then possible to calculate an average busy-hour call holding time and construct a matrix of forecast busy-hour traffic flows in erlangs or CCS, both within the private network and between this and the public network(s). Typically, busy-hour traffic is found to be about one-quarter to one-fifth of the total traffic over an average working day. The process is illustrated in the following simple example for the PBX shown in Fig. 6.1, where the network is considered to have a single network-wide busy hour:

Average busy-hour call holding time: 3 min = 0.05 h or 1.8 hundred seconds
Average busy-hour extension-to-extension calling rate: 0.9 calls/h
Average busy-hour extension-to-public-network calling rate: 0.35 calls/h
Average busy-hour public-network-to-extension calling rate: 0.75 calls/h

Traffic flows:

Extensions to extensions: $0.9 \times 100 \times 0.05 = 4.5$ E
                        or $0.9 \times 100 \times 1.8 = 162$ CCS
Extensions to public network: $0.35 \times 100 \times 0.05 = 1.75$ E
                        or $0.35 \times 100 \times 1.8 = 63$ CCS
Public network to extensions: $0.75 \times 100 \times 0.05 = 3.75$ E
                        or $0.75 \times 100 \times 1.8 = 135$ CCS

The collected information relative to each switching centre will also indicate the number of *successful* call attempts that must be handled by the call-control processor during the busy-hour; successful is emphasised here because it is this figure that will be obtained from enquiries of users. In practice, there will always be some additional, unsuccessful call attempts (which would be included in any measured figures from an established network) for which an allowance must be made in the determination of equipment quantities. It has to be remembered that the average call holding time used to calculate traffic quantities will also have been derived from a sample of successful calls; this figure cannot, therefore, be used in conjunction with subsequent measured call counts, which will require the use of an average value derived from a mix of successful and unsuccessful call attempts.

Advances in technology can be expected to continue to enable ever more complex circuitry to be realised more cheaply in less space. Competition between public-network operators is, however, likely to direct any savings made as a result of lower transmission and switching costs towards a reduction of tariffs rather than the dimensioning of these networks on a more liberal basis. The need for detailed traffic measurement and forecasting processes for public networks can, therefore, be expected to continue into the foreseeable future. The experience in traffic measurement and forecasting built up over the years for both public and private telephone networks has been derived from a situation where the traffic on these networks has consisted almost entirely of telephone calls. This situation is now changing; although telephone traffic will still predominate, public telephone networks are being progressively transformed into integrated-service networks (the ISDN: see Section 4.1) and private networks are increasingly being designed to carry integrated speech and data traffic. Some types of traffic will be confined to part of a network, for example packet-type data which is routed over the periphery of the ISDN to gain access to a separate packet-switched network. Packet-type data traffic is not suited to traffic-flow measurement on the traditional erlang or CCS basis, yet it will be carried over the same circuits as call-based telephone traffic. The current, long-established methods for measuring and forecasting traffic on circuit-switched telephone networks will no longer be appropriate for universal application in these changing circumstances, and new practices can be expected to evolve which take advantage of improved facilities for the automatic extraction of data and the, increased use of computerised data analysis and forecasting techniques.

6.3 Message-switched data-network traffic measurement and forecasting

The messages transmitted over a public message-switched store-and-forward telex network are of varying length, ranging from a simple, very short, answerback signal to several hundred words of a user-to-user message. In this respect such message-switched networks are akin to circuit-switched networks, and message holding time is an essential ingredient of traffic-flow measurement. Messages are considered in the same way as telephone calls; the sum of message holding times over one hour will provide erlang or CCS figures, and the calculation of an *average message holding time* from a representative sample of messages enables average traffic-flow figures to be directly computed from a message count. The switching centres in a public message-switched store-and-forward network, therefore, provide information on congestion and the number of message attempts and successfully transferred messages, in much the same way as those in a circuit-switched telephone network.

The *store-and-forward* feature introduces a new factor that must be taken into account in traffic measurement. Messages are stored before onward transmission, the storage acting as a buffer during peak traffic periods (Section 3.2). The availability of this storage space means that traffic routes do not need to have reserve capacity for the peak traffic demand to the same extent that they do in a circuit-switched network and can, in consequence, be designed to be more heavily loaded under normal traffic conditions. In other words, circuits in a network with a store-and-forward capability can be operated at a higher occupancy and hence a *poorer grade of service* than circuits in a network without this feature. This means that fewer circuits are needed to carry a given amount of traffic than in an equivalent situation in a circuit-switched network. As a result of this improved efficiency, however, the users experience increasing network delay when the available circuit capacity is fully employed during peak traffic periods and, should the storage capacity be fully taken up, are temporarily unable to transmit new messages. An economic balance has to be struck between switching and transmission equipment provision, storage capacity, the tariff to be charged and an average network delay that is tolerable to users (Section 6.5). Details of the current state of network delay are, therefore, an important ingredient in the management of an established message-switched store-and-forward network. Network delay is the sum of switching-centre message handling times and link transmission times; the latter are known, fixed quantities for any particular network but message handling times vary according to demand and must be measured and recorded during the busy hour at each switching centre.

The forecasting techniques applied to a *public* message-switched store-and-forward telex network, being based on a similar approach to traffic measurement, follow the principles already outlined for a circuit-switched telephone network. Traffic forecasts in erlangs or CCS are produced for individual routes and switching centres, each element of the network being augmented separately as and when required. A *private* network makes use of public-network-type switching equipment, but traffic measurement is not necessarily carried out to the same extent as in a public network. Forecasting techniques are often more closely allied to computing rathèr than telecommunications practice, forecasts being constructed solely on the basis of terminal equipment needs. Changes to the number and type of terminals may alter the nature and amount of traffic to be carried; this can have a significant effect in a small private network, whereas the effect would be lost in a large public network. However, this problem is to some extent counterbalanced by the fact that a single user organisation only is involved, simplifying the collection of information on future needs. Forecasts for both existing and new networks are constructed from estimates of data message flows between stations or areas of common interest at the busiest period of a working day. Information is collected in terms of the number of individual *messages* and *average message duration* (this compares with unswitched computing networks, where traffic is measured in terms of message pairs because each message to a computer automatically invokes a response; duration is similarly in terms of an average message pair). These figures are then used directly to forecast (a) circuit requirements on the basis of per-circuit *occupancy* without recourse to traffic-flow calculations in erlangs or CCS, and (b) the maximum message handling rate required of a switching centre.

6.4 Packet-Switched data-network traffic measurement and forecasting

As might be expected from the message-switched store-and-forward ancestry of packet switching, there are similarities in the traffic problems encountered in both types of network. There is, however, one very important difference: instead of variable-length messages, sometimes requiring many minutes to transmit, the packet-switched network carries short packets of fixed maximum length which are, on the interconnecting links, extended to slightly longer but still fixed-length frames (Section 3.3). The fixed length of packets and frames renders holding-time measurements unnecessary; each link operates at a fixed transmission speed and is able to carry a fixed maximum number of packets/frames per second. Measurements in terms of traffic-flow units are similarly unnecessary; for a given circuit occupancy, the number of circuits required on a route can be determined directly from a forecast of the number of packets per second that need to be transferred over that link.

Another difference between packet-switched and message-switched networks arises from the *shortness* of packet holding time – very much shorter than a typical average message holding time. This has the effect in a packet-switched network of reducing (a) cross-network delay and (b) the amount of network storage needed, because the packets can be interleaved on a circuit much more efficiently than longer variable-length messages. The aim in a public packet-switched network is to keep user-to-user data-transfer times sufficiently short for interactive operation with a computer, requiring delays of no more than about a third of a second. If delays of this order are to be achieved and maintained, the accurate measurement of handling times at each packet handling point is essential. Evidence of increasing network delay obtained from such measurements enables network bottlenecks, resulting from network design deficiencies or unforeseen traffic growth, to be pinpointed.

Private packet-switched data networks, especially when based on the local-area-network concept, are again smaller than their public equivalent, with less comprehensive traffic measurement facilities. However, the nature of packet-type data traffic is such that peak circuit usage in a small private network can be quantified more accurately than in a message or circuit-switched network of equivalent size. Moreover, as with a message-switched store-and-forward network, traffic demand in excess of the designed network capacity initially results in increased user-to-user response times; the network is only blocked to new traffic when it is severely overloaded. There is no doubt that even large private packet-switched networks can be operated effectively without traffic measurement, but where measurements are taken on a regular basis they provide a much better understanding of how a network is performing. The deterioration of network delay can more easily be detected and remedied before it can cause serious problems. In addition, especially for the largest networks, a degree of economic optimisation can be achieved that otherwise would not be possible.

Forecasting for a *public* packet-switched network is a very different process from that used for circuit- and message-switched networks, although a forecast for an established network will be similarly derived from a historically based trend line. Measurements are in terms of *packet handling time*, and *throughput* in *packets per second*. Measured packet handling times are compared with a target figure, and this is often the sole criterion for deciding if network augmentation is required. Augmentation usually has to be considered on a network-wide rather than a piecemeal basis because of the greater interdependence of network components. The alternative routing capability, whereby packets are not constrained to a single path between source and destination, means that traffic flow on the individual traffic routes cannot readily be predicted with the degree of accuracy possible for other types of switched network. A worthwhile analysis of traffic flows would prove to be extremely complex, if not impossible, in these circumstances, and any attempt at analysis would be difficult to justify economically. The forecast total network throughput in packets per second is, therefore, usually simplistically apportioned throughout the network on the basis of the demand on each switching centre, the circuits being deliberately underloaded as a precaution against forecasting and apportionment error (Section 6.5).

A forecast for a *private* packet-switched network is likely to be based solely on a periodic reassessment of the capability and usage of each station, as for a private message-switched network. In simple terms, the forecast network throughput in packets per second is obtained by taking into account the output from stations which are expected to be transmitting at the same time during the busiest period of a working day ('the same time' is intended loosely here since stations transmit one packet at a time, the packets from different stations being interleaved). One station transmitting implies another station receiving; hence it might appear that no more than half the total number of stations can be transmitting at the same time. However, the store-and-forward facility means that, in practice, data transmitted from more than half the total number of stations may be present on the network. In addition, the possible need for the multiple addressing of messages, and thus packets, from a single source to a number of destinations has to be taken into account.

Forecasting for a *new* public packet-switched network is made more difficult by the fact that it is possible to survey only a sample of the potential market, and any information available on typical circuit usage worldwide may prove not to be representative for the country concerned. This problem is not so acute for a new private network, because a full survey of the market is possible within the single organisation involved and information on potential network usage can be gathered in much greater detail. As always for complex networks which may require more than one transmission path, the forecasts must be constructed initially in terms of data flows between areas of common interest rather than individual stations, to permit full consideration to be given to alternative network solutions.

6.5 From forecast to equipment quantity: the effect of grade of service, circuit occupancy and network delay

The traffic measurement and forecasting procedures described so far in this chapter culminate in the provision of a forecast (or series of forecasts) of future telecommunications demand during the busy hour for a particular network or network component. Whether the requirement is to establish a new network or to augment one that already exists, the ultimate objective is to equip that network to carry the amount of traffic forecast at an economic cost to the network provider whilst offering an acceptable quality of service to its users. The next and final stage in this process, therefore, is the conversion of a traffic-flow forecast into an appropriate quantity of equipment. Once again, the procedure and techniques used differ according to the type of network and switching principle concerned.

The conceptual design of *public networks* is undertaken on the basis of long-term forecasts. Different network configurations, traffic routing techniques, etc. are considered from both the technical and economic points of view. This is not a once-and-for-all process, however; the history of telecommunications shows that, as the pace of change continues to quicken, each new generation of equipment becomes obsolete in less time than its predecessor. Technological advances, together with long-term changes in traffic characteristics and increased customer expectations, necessitate the reconsideration of at least some aspects of network development every few years. The end result of these continuing deliberations is a set of occasionally changing ground rules to control network augmentation and reconfiguration, and the replacement of obsolete or unreliable plant. Short- to medium-term forecasts are then prepared and equipment quantities calculated as separate ongoing periodic tasks, although it is essential that there should be sufficient flexibility in both processes to allow rapid expedient action to be taken where unforeseen changes in demand warrant this.

Businesses that operate a large *private network* usually employ a qualified telecommunications manager to ensure that the network provides an efficient service and continues to meet the changing needs of the organisation. One of the tasks of the manager is periodically to consider – either directly or using the services of a consultant – the long-term development of the network, on a similar basis to that described for public networks. Small private networks, however, are seldom professionally managed; they are designed initially by the equipment manufacturer or an independent consultant, and advice on augmentation, limited updating or complete system replacement is sought only as and when deemed necessary by the user organisation. The tasks of forecasting, network design and/or system choice where required, and the determination of equipment quantities, are then usually carried out as an integrated but isolated operation. Network development does not, therefore, receive the benefit of continuing reappraisal but proceeds in a series of disjointed steps.

For private networks, it is necessary to differentiate between cases where (a) the network configuration and system are predetermined (for example, where an existing network is to be augmented or a new network equipped with a particular type of PBX or LAN), and (b) an optimum solution to a particular communications need is required. In case (a) the forecast will be tailored to the particular network/system concerned and the conversion of forecast requirements to equipment quantities is a relatively straightforward task. A more general forecast is appropriate for (b) and a higher level of expertise is demanded by the need to make a number of optimum choices from a range of network/system options before equipment quantities can be determined.

6.5.1 *Telephone and telex networks*

Similar measurement and forecasting processes are used for public telephone and telex networks because in both cases traffic is presented to the network in a random manner and carried as variable-length calls or messages. A traffic-flow forecast is based on two different variables – the number and duration or holding time of calls/messages. The number of busy-hour calls/messages used in the forecast is the average of a representative sample of measurements taken during past busy hours and extrapolated into the future; in practice, the fact that calls/messages arrive in a random manner means that the actual number of arrivals during any particular busy hour may differ from the forecast average number. Similarly, the call/message duration or holding time used in the forecast is the average of a representative sample of measured holding times which also, in practice, differ in length in a random manner. The forecast number of traffic-flow units is, therefore, the product of two *average* values of random variables.

Any average value implies that although a proportion of the real values concerned will be at or close to that average, some real values will be significantly higher or lower. The first step in converting a traffic-flow figure to a quantity of equipment, therefore, is to determine how real values are likely to be disposed about the two different averages in the case of call/message-based telecommunications traffic and what the upper and lower limits are likely to be. In any such situation, the probable distribution of real values about an average value can be represented by a particular mathematically based curve, permitting calculations to be made which take into account the likely spread of real values. This requires certain limiting assumptions to be made about the nature of the traffic under consideration; in mathematical terms, the arrival of calls/messages can then be represented by the *Poisson distribution* and the duration of calls/messages by the *negative exponential distribution*. On this basis, formulae are derived which allow the number of circuits required for a particular level of traffic flow to be calculated at a given *grade of service* (GOS). Although a traffic route is designed to a particular GOS at the normally expected busy-hour load, the actual quality of service experienced by users deteriorates when the route is overloaded. The variability of traffic flow from hour to hour and day to day, together with the fact that these formulae are based upon more than one level of probability, make it inevitable that there will be occasions when congestion occurs as a result of localised traffic levels exceeding the designed network capacity.

The larger the number of calls to be carried on a particular traffic route, the more efficiently will they be interposed over the larger number of circuits involved; the traffic is said to be *smoothed*, and each circuit is thus able to carry more traffic at the same GOS as route size increases. However, the economic advantage offered by the improvement in per-circuit traffic-carrying efficiency with route size is to some extent counterbalanced by the fact that the GOS will then deteriorate more rapidly if the route is overloaded. Formulae can be designed to take account of this by allowing for more circuits than would be justified by the required normal-load GOS as route size increases, such that the actual GOS experienced will not deteriorate below a stated value for a given percentage of overload traffic. The use of such formulae is particularly advantageous in public networks where routes may consist of hundreds or even thousands of circuits, but they are of less relevance to a small cost-optimised private network.

The tedious and repetitive use of complex formulae is avoided by producing sets of look-up tables which relate erlangs or CCS to numbers of circuits for a particular set of circumstances. A table which takes into account a safety GOS limit at typically 10% and/or 20% overload, as well as the required normal-load GOS, is referred to as a two-criteria or three-criteria table. An example of a simple normal-load *traffic table* which covers six different GOS values is included at Appendix A. Reference to this table shows that, for forecast traffic flows of 5 E and 15 E, circuit requirements (number of circuits) are as follows:

Forecast traffic flows	*Grade of service*: 1 lost call in					
	10	20	33	50	100	1000
	(0.1)	(0.05)	(0.03)	(0.02)	(0.01)	(0.001)
5 E	8	9	10	10	11	14
15 E	18	20	22	23	24	28

These figures illustrate how (a) GOS is improved by adding more circuits and (b) traffic carried per circuit increases with circuit-group size.

The term *full availability* in the heading of Appendix A means that it has been compiled on the assumption that each call attempt requiring the use of one of the circuits of the group to be dimensioned will have access to *every* circuit of that group. Modern electronic switching systems, both analogue and digital, generally offer full availability; hence no other alternative need be considered. The need to define availability arose because of the physical limitations of electromechanical switching equipment, which usually allowed only *limited availability* in the sense that each call attempt had access to only some of the total possible circuit outlets; an availability of 20 means that each call attempt has access to 20 possible circuit outlets irrespective of how many more circuits there are on a route in total.

GOS has so far been considered in relation to one particular stage in a chain of switched stages between two network users. Each stage is dimensioned to a particular GOS standard, according to its position in the network. A GOS of 0.01 means that, if traffic flow is at the forecast level used in the dimensioning process, the probability is that 1 busy-hour call attempt in every 100 will fail at that particular *stage* of a call attempt because of a lack of free circuits. The best possible GOS is zero, which implies that all call attempts within the designed busy-hour capacity should find a free circuit awaiting them at that stage; a route is then said to be *fully provided*.

What matters to the network user, however, is the degree of success experienced when attempting to make a call during the busy hour, in the process of which a *number* of such stages is encountered in succession. This *overall* or *end-to-end GOS* is influenced by (a) the number of stages encountered in making the call attempt, (b) the timing of the call attempt in relation to the busy hour(s) (possibly slightly different at each stage), (c) the GOS standard applied to dimension each stage, and (d) the actual level of traffic flow at each stage relative to that forecast. The first three of these points should be self-evident, but the last requires some amplification. Circuits are connected or equipment installed to carry the expected busy-hour level of traffic flow at a forecast date, typically from 1 to 3 years ahead; this timescale takes into account factors such as the growth rates being experienced, forecasting procedures, equipment modularity, time between ordering and installation, etc. It follows that a designed GOS standard at a particular stage will apply only if and when the actual traffic flow *attains* that forecast level. If the forecast level and timing are correct, this will happen at or near the forecast date. A difference between the actual and forecast traffic levels may be due to the inaccuracy of the forecast or may arise because the date at which the forecast is intended to apply has not yet been reached. In the latter case, assuming a growing network and an accurate forecast, the actual traffic level will be less than that forecast. It will be seen from Fig. 6.2 that the

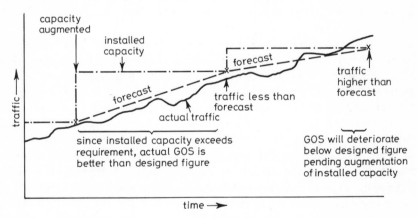

Fig. 6.2 *Relationship between forecast, installed capacity and actual traffic levels*

overall GOS experienced by callers during the network busy hour should be *better* than the designed standard in the interim between the date of augmentation and forecast date of the traffic-flow level used to dimension that augmentation. Irrespective of the date of augmentation relative to a call attempt being made, callers should also experience a better overall GOS outside the network busy hour than during the busy hour. It will also be apparent from Fig. 6.2 that when the forecast level of traffic flow is attained, further augmentation will be required if the GOS is not subsequently to deteriorate below the required standard.

The actual overall GOS likely to be experienced by a user on any particular call is virtually impossible to predict; even a generalised estimate involves calculations at a great many levels of probability. It is not surprising, therefore, that the present-day GOS standards for public networks – more stringent than those generally applied to private networks – have evolved largely on a trial-and-error basis with little economic justification. GOS standards do need to be adjusted to take full economic advantage of the various possible methods of routing traffic (see Sections 7.5 and 8.1), but apart from this they tend to remain unchanged for many years. The quality of service offered by public networks, however, has and still is steadily improving as a result of better equipment reliability, and many networks are in the process of adjusting to commercial competition for the first time in their history. These changes bring into economic question the traditional practice of engineering a network to a set of fixed GOS standards. Essentially changes to public-network GOS standards would result in changes in the balance between call charges and congestion; the economics of such changes, taking into account the likely user reaction in a competitive environment, have yet to be fully investigated, but a more flexible approach to GOS standards could be to the ultimate financial benefit of both network operators and users. The economic significance of fixed versus flexible GOS standards for private networks, however, is of less relevance because of smaller network size.

For a store-and-forward message-switched network such as the public telex network, overall *message transit time* – directly influenced by the average message handling time at each switching centre – cannot be allowed to exceed an acceptable level. A network that is less heavily loaded and, therefore, less efficient in terms of its traffic-carrying capacity, will incur shorter network delay; thus transit time can only be reduced at the cost of providing more circuits and switching-centre equipment. A balance has to be struck between network costs and delay. The causes of excessive delay are complex to evaluate in precise terms; not only does the amount of delay experienced depend on the inter-relationship between traffic-route capacity, switching-centre capacity and storage capacity, but delay is inversely related to the rejection of messages during busy periods. If it is assumed that traffic-route and switching-centre capacities are correctly matched to the actual traffic flows, their effect on delay can be isolated. The effect of varying the storage capacity is then more easily understood; insufficient storage actually reduces delay but causes more messages to be rejected during busy periods, whereas too much storage increases delay but reduces message rejection. Within the limits imposed by the installed storage capacity, therefore, the amount of delay can be broadly controlled by the number of circuits provided on traffic routes and the amount of switching equipment at switching centres – in other words, by the GOS standards which are applied.

6.5.2 *Data networks*

The criteria for deciding circuit and equipment quantities for a true data network are entirely different from those so far described. Neither forecasts in terms of traffic-flow units nor the concept of GOS are used. Even the terminology is different: the telecommunications engineer speaks of *dimensioning* a network from a traffic forecast, whereas the private-network data communications engineer follows computing practice in *sizing* a network from a forecast of data throughput. Forecasts of peak-hour data throughput for any private network in which the data is message based rather than packet based may be expressed – on a link-by-link basis – in terms of an average number of messages (message-switched networks) or message pairs (unswitched networks) per hour; it is then necessary to estimate, for a given transmission speed, an average message or message-pair duration. Alternatively, the forecast for an unswitched network may be expressed directly in bits per second.

Suppose, for example, that the average peak-hour throughput on a particular unswitched data-network link is forecast to be 145 message pairs per hour, each message pair being of average duration 65 seconds at a transmission speed of 2400 kbit/s. A simple calculation shows this to equate to an average data throughput of $(145 \times 65/3600) \times 2400 = 6.3$ kbit/s. This would require three 2400 bit/s leased circuits, giving a capacity of 7.2 kbit/s and an average circuit occupancy of $6.3/7.2 \simeq 0.9$. However, by increasing the speed of transmission over the link to the next available leased-circuit transmission speed in excess of the requirement – that is, 9.6 kbit/s – one circuit only is needed for the link and the occupancy is reduced to $6.3/9.6 \simeq 0.7$, allowing a greater margin for forecasting error. It is then worth questioning what the effect would be of using a leased circuit operating at the next available lower transmission speed, namely 4.8 kbit/s, less than the requirement figure of 6.3 kbit/s. Clearly, not all the required traffic could be carried during the peak period and the transmission of some messages would be delayed. However, where such delays are acceptable – especially where there is a sufficient proportion of non-urgent messages which can be administratively withheld until demand has slackened – the lower-speed circuit might offer a more cost-effective solution. The network designer, as always, has to make a choice within the limitations imposed by the particular operational requirements of the network, the cost and technical capabilities of possible network solutions and modems, terminals, etc., and the leasing costs of the available range of private circuits. Circuit requirements can be determined and transmission speeds optimised in this manner for each link separately. One important point is demonstrated by this example: the enormous difference in physical size between a public telephone network, where the number of circuits required for a traffic route may well run into thousands, and the typical private data network, where concern is often over the capacity of a single circuit.

Packet-switched networks present a more complex problem to the network designer, if only because of the wide choice of networks and systems available. For a network equipped with one or more switching centres of the type used for a public network, however, where network delay can be determined from switching-centre packet handling time(s), it is possible to calculate the circuit requirements for a link independently. Forecasts of peak-hour data throughput are in terms of an average number of packets per second, where the fixed frame bit content is a known quantity for a given type of network. Suppose, for example, that the average peak-hour data throughput on a particular packet-switched data link is forecast to be 30 packets (and therefore frames) per second, and each transmitted frame consists of 140 octets $= 140 \times 8$ bits. This means that the link is required to transmit an average of $30 \times 140 \times 8 = 33.6$ kbit of data per second.

Here again, the network designer has a choice; a 56 kbit/s digital leased circuit loaded with a data throughput of 33.6 kbit/s would be utilised at an occupancy of $33.6/56 = 0.6$, and a 48 kbit/s circuit at an occupancy of $33.6/48 = 0.7$. The next lower transmission speed available over a public-network leased circuit is 19.2 kbit/s; it is clear, even without a detailed study, that the capacity of such a circuit would be so much less than the demand placed upon it than an excessive buildup of network delay would inevitably result. Either a 56 or a 48 kbit/s circuit could carry the required data throughput, the higher-speed circuit operating less efficiently but offering a greater margin of safety against overload. The choice between the two possible circuits, taking into account overall network performance, is once again a matter of cost/benefit analysis.

This example and that relating to a message-based data link illustrate the general principle for determining circuit capacity on an independent-link basis, but neither takes any account of *delay*, which is further discussed in Section 8.2. For many private packet-switched networks, however, the link-by-link approach to sizing is not applicable. For example, a baseband local area network (LAN) has its switching facility distributed among the stations situated along a single, shared transmission path (Section 5.3), and each manufacturer's variant on the general principle has its own particular finite capacity in terms of (a) data throughput in packets per second and (b) the number of stations that may be connected to it. For these networks the problem is not to determine the capacity needed on the transmission path; instead, the network designer is concerned that the fixed capacity of a particular transmission path will suffice to meet the demand which is forecast to be placed upon it without introducing unacceptable network delay. Network delay cannot, therefore, be considered as a separate issue. The sizing calculations for a given network delay are complex, as are all calculations based upon the probability of delay attaining a certain figure, and manufacturers commonly employ a computer program to provide rapid solutions for their particular proprietary system(s).

6.5.3 Network design
Network design, which incorporates network/system choice, is a complex task that requires a detailed knowledge of the networking principles involved and proprietary systems available. The telecommunications engineer is generally

limited to a single networking principle but must exercise a choice on such matters as switching-centre locations and sizes, full or partial interconnection, traffic routing methods, etc. The data communications engineer must choose from a wide selection of networking principles to suit different needs. Both have to decide between alternative means of providing the interconnecting links and competing proprietary systems. Both have ultimately to dimension or size their networks.

Public-network design is a step-by-step process involving a number of different areas of expertise, and the engineers concerned tend to specialise in one particular task. Most private-network design engineers will have acquired their knowledge in either telecommunications or computing, seldom both, and their experience also may not extend across the whole of these very wide areas of expertise. Since private-network design is most likely to be undertaken by a small integrated team or even one individual, it is important for the manager wishing to install a new telecommunications facility to choose a design organisation with a proven record of experience and success. A manager cannot hope fully to understand the problems involved without extensive study but, by concentrating on a few general traffic-related design and dimensioning/sizing considerations, can gain a sufficient understanding of the subject to question decisions intelligently. In the next two chapters, therefore, the effect of these traffic considerations upon switching system and network performance respectively is discussed in greater depth.

Chapter 7

Traffic considerations in switching-system choice and dimensioning

7.1 What is a switching system?

A telecommunications switching centre is constructed from an integrated range of equipment modules collectively termed a *switching system*. The task of deciding which and how many of these modules are needed for a particular application, as outlined in Chapter 6, is generally referred to as switching-centre *dimensioning*. The dimensioning process is, however, constrained by the pattern and extent of the breakdown into equipment modules – in other words, by the overall conceptual *design* of the switching system itself. A switching system is designed for the cost-effective handling of one particular type of traffic; a system intended for telephone and other call-based traffic is significantly different from those for message- or packet-based data traffic.

Local area networks (LANs), although they have a packet-switching function, are classed as network systems rather than switching systems; the equipment from which they are constructed is, therefore, discussed separately in Chapter 8.

A particular switching system is usually supplied exclusively by one manufacturer but exceptionally, where the basic design work has been undertaken by or in cooperation with a network operator, several manufacturers may supply what is, to all intents and purposes, the same product. Private-network switching systems may be variants of public-network systems or, in the case of telephone networks, PBXs which are designed specifically for their purpose. A switching system catering for all sizes of switching centre would not be an economic proposition; hence all switching systems cover a limited capacity range, with both upper and lower limits. At the upper end of the scale, a large modular switching system for a public telephone network may cater for from 2000 to perhaps 60 000 local circuits, whereas a system designed for small PBXs will have a very restricted size range. In general, a switching system used at or close to the lower limit of its range is less cost-effective, in terms of initial cost per user and per unit of traffic handled, than when it is used in its middle and upper ranges.

All switching centres, and the switching systems from which they are constructed, can be broken down conceptually into a number of major functional elements, as already demonstrated for a PBX in Section 5.2. Each element has a set of clearly defined tasks to perform. A switching system is also broken down physically into a number of equipment modules, which do not necessarily coincide with the functional breakdown; an element may be part of an equipment module or, at the other extreme, it may be built up from a large number of equipment modules. The tasks, overall, are broadly similar for all switching systems designed to carry the same type of traffic, but the circuitry which performs those tasks and the breakdown into equipment modules can be realised in many different ways. As with any other types of manufactured product which are designed for a particular purpose, no two switching systems are exactly alike. Each is designed as a complete product range of equipment suitable for switching centres of a given type and size range, and its modular concept, control software and total electronic/physical realisation are unique to that system. Where a system is made by more than one manufacturer, each may produce a variant which differs in its detail. The equipment modules of a particular switching system cannot, therefore, be interchanged with those of other switching systems (this restriction does not necessarily extend to ancillary items, some of which may be available from independent sources and suited to a range of different switching systems).

The fact that a switching-system design is being manufactured does not necessarily mean that it is well designed. The system may be good in some respects and poor in others, well suited to some applications and not others, as with any other manufactured product. Some of the factors which influence system choice – particularly those which affect running costs such as reliability, ease of maintenance and power consumption – are outside the scope of this book. There are, however, certain common capacity-related considerations which make some switching systems more suitable than others for a particular application and have an effect on the cost of initial purchase and subsequent extension, as discussed in the remainder of this chapter. Different types of traffic give rise to different capacity-related issues, covered separately in the next two sections. The issues concerning processor control and traffic routing are not necessarily confined to particular types of traffic, however, and these two areas of switching-system design are dealt with in subsequent sections. A final section discusses the considerations which apply particularly to the switching of integrated speech and data traffic.

7.2 Telephone switching systems and PBXs

Each local switching centre or PBX is located at the hub of a network of lines to public-network users or private-network extensions and, in large networks, there will also be links to other switching centres. These links take the form of shared traffic routes in a public or large private network, whereas a PBX accesses a public network over individual circuits and other PBXs over separate tielines. The major functional elements of any telephone-network switching centre of this type are broadly the same as those discussed in Section 5.2 for a PBX, but only some of these are likely to have a significant effect on costs. A more general version of this breakdown is, therefore, appropriate to the following discussion, as shown in Fig. 7.1; it must be stressed, however, that this is typical only and is *not* representative of all switching systems. The principal functional elements which may influence cost, as shown here, are (a) the switch, (b) line circuits or, for a PBX, extension circuits, (c) own-switching-centre connecting circuits, (d) traffic-route or equivalent PBX terminating circuits and (e) pool signalling circuits. The control element is, of course, equally important but is considered separately in Section 7.4.

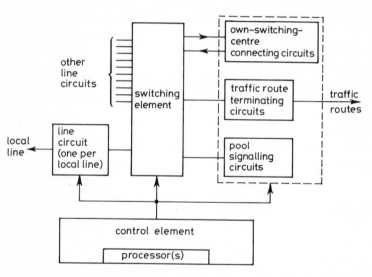

Fig. 7.1 *Principal cost-influencing functional elements of a telephone switching centre*

The *switching element* of a switching centre or PBX is commonly referred to as a *switchblock* and typically consists of several successive switching stages, each having a number of switching points. The number of switching stages and switching points per stage varies according to the capacity of the switchblock and thus of the switching system. The switchblock is shown in outline in Fig. 7.1 as having two 'sides', connection points on the left being switched as required to link with those on the right. An analogue-system switchblock has a physically separated connection point per circuit, but each connection point in a digital system terminates a 2 Mbit/s or 1.5 Mbit/s stream within which 30 or 24 circuits are separated in time (Section 2.8). The commonly employed switchblock arrangement shown requires users/extensions of a local switching centre or PBX to be connected to the left-hand side only. Traffic-route circuits or their equivalent, together with various ancillary circuits such as those which link own-switching-centre calls and provide connections to signalling sender/receivers, tone and ringing supplies, etc. (see also Section 5.2 and Fig. 5.3 for a typical PBX), are connected to the right-hand side. The depiction of a switchblock as having two sides is strictly only valid for a small switching system; in larger systems it may not be necessary to use all switching stages to make a particular switched connection, and economies are achieved by terminating some traffic-route/ancillary circuits between stages.

The switchblock in modern switching systems is so designed that a call attempt cannot fail because of a shortage of connecting paths across it, and it is said to be internally *non-blocking*. However, because all users do not make or receive a call at the same time, it has more external connection points for users than for traffic-route/ancillary circuits and acts as a *traffic concentrator*. The number of connection points on the left is determined by the maximum number of local circuits or extensions to be catered for by that particular switching system. The number required of each type of circuit on the right is determined by a forecast of either traffic flow or call attempts, as appropriate.

It will be seen that the speech path between two users, both of whom are connected to the same switchblock, is routed through the switchblock twice – once in each direction. For an *outgoing* call from a user on the switchblock to another user elsewhere or for an *incoming* call to that user from elsewhere, the speech path is routed only once through the switchblock. Additional temporary paths across the switchblock have to be established during call set-up to associate the required ancillary circuits. Similar switchblock patterns obtain for public-network trunk/toll switching centres designed exclusively for tandem switching between other switching centres but, since there are no directly connected users, all external connections are traffic routes. The traffic routes from local switching centres then take the place of users.

The switchblock of a small switching system may be supplied as a single invariable entity, whereas all larger systems have a modular switchblock structure. A modular switchblock has a minimum size dictated by the size of a single module and can be extended to its maximum capacity in specific steps, by the addition of further modules. The switchblock is, therefore, a potential limiting factor on the ultimate capacity of a switching centre. In addition, although the switchblock is not a particularly costly element of a modern switching centre or PBX, module size can have an important influence on extension cost and timing.

A *line* or *extension circuit* provides an interface for signalling and transmission to and from a user over a local line, converting between the higher line-power/ringing voltages on the line and those used in the switching-centre electronic circuitry. These circuits are not traffic dependent, being required on the basis of *one for each line*. A small PBX may be equipped with a fixed number of line or extension circuits, but in most cases they are supplied as multicircuit equipment modules. Miniaturisation is not yet possible to the same extent as in most other elements of a switching centre because of the higher-voltage circuitry required; hence module sizes are usually in terms of no more than a few circuits, and *per-circuit costs* can be relatively high. This is an important factor in system choice, especially where growth in the number of users is anticipated. Provided the line- or extension-circuit modules are quickly obtainable on a one-off basis and easy to install, relatively short-term forecasts can be used to determine the number required. It is then possible to reduce costs by fitting only the minimum number of modules consistent with the expected requirement at any given time, in the knowledge that additional modules can be obtained at short notice to meet unforeseen demand.

When choosing a switching system, another consideration with regard to line or extension circuits is the maximum capacity limit on the number of users that can be served. In a small system this is normally fixed physically by the space allotted for line or extension circuits, as determined by the switching-system designer. This advertised limit will have been matched to the traffic-handling limitations of the remaining elements of the system, on the basis that each user generates a certain average amount of traffic in the busy hour. Although the number of line circuits in a public-network-type switching centre is not physically limited in this way, their number will still be restricted by the traffic-handling capacity of other elements of the system on which there *is* a physical limitation. Call-handling capacity is normally specified by the manufacturer in terms of a maximum *busy-hour call-attempt* (BHCA) rate, which equates to a particular average calling-rate limit per user's line if all lines are connected. An average per-line calling rate of 0.5 calls/hour, for example, represents one call of average duration every other busy hour. If a high proportion of users generates traffic in excess of this average limit and the advertised maximum number of line circuits is in use, the system will be overloaded. The number of users that can be served, therefore, may be *less* than the manufacturer's advertised limit in particular circumstances. The actual maximum number of users that can be served by a particular switching centre is calculated during the dimensioning process, from the forecast average calling rates.

A special connecting circuit, which may be known by various (sometimes system-specific) names, is needed to link two users on the same switching centre. Here the descriptive title *own-switching-centre connecting circuit* is used. A path is set up through the switchblock from the calling user to one of a pool of these circuits, and then back through the switchblock to the called user. The number of circuits in the pool is calculated on a grade-of-service basis from a forecast of own-switching-centre traffic which, in PBX terms, means the amount of internal traffic between extensions. The main problem here is one of possible forecasting inaccuracy combined with the fact that an equipment module usually provides only a few circuits. An adequate reserve of circuits against forecasting error is particularly necessary for a new switching centre or PBX where traffic measurements are not available as a guide to the forecaster. Own-switching-centre connecting circuits are not particularly complex and the cost of an extra module or two is low compared with the possible cost that could be incurred as a result of congestion.

Traffic-route or *equivalent PBX circuits* are required at both analogue and digital switching centres to terminate four-wire analogue transmission-system channels and two-wire 'physical' lines. Their purpose is to provide (a) four-wire to two-wire conversion where required and (b) the necessary electrical conditions for signalling over the channel or line on a channel-associated basis. Where neither of these facilities is required – for example, when digital channels are terminated and there is no need to use analogue-type channel-associated signalling – terminating circuits on a per-circuit basis are unnecessary. Terminating circuits thus constitute part of the private-network switching equipment in all cases where analogue public-network leased lines are employed but, where the leased circuits are digital, advantage can be taken of their inherent signalling capability or common-channel signalling as in the public network.

Switching-centre termination circuits vary considerably in complexity and cost, depending on the type of signalling and connection involved. Generally, those required for the longer-distance analogue transmission routes will be the most expensive since each not only incorporates a four-wire to two-wire conversion device but also requires more complex signalling arrangements. Such circuits are likely to be supplied as single-circuit modules, whereas others of less complexity may be supplied as modules of more than one circuit. Each termination must be matched to a circuit; hence terminating-circuit numbers are forecast on the same route-by-route basis as the linking circuits. Unless a single PBX only is involved, the individual switching-centre designer usually has no say in the choice of GOS standards since these are determined on a network-wide basis as a matter of network dimensioning policy; network GOS standards are discussed in the next chapter.

The last type of circuit to be considered is the *pool signalling circuit*. This title should not be confused with the per-circuit signalling circuits discussed in the previous two paragraphs. Pool signalling circuits are provided as an independent group not directly associated with any particular traffic route or user's local line, but accessible via the switchblock to any call attempt on demand. Voice-frequency (VF) and multifrequency (MF) sender/receivers are typical of such pool circuits. Either may be required for public-network circuits which employ VF or MF signalling, and versions of the latter are needed for MF signalling over local lines where users are equipped with fast-signalling keyphones, as commonly supplied with modern PBXs. These circuits are relatively costly and, since they are needed only during call set-up, it is more economic to select one from a small pool as and when necessary than to provide the much larger number that would be needed on a per-line basis. Some other so-called peripheral circuits are also provided on a pool basis – for example, those circuits which generate tone and ringing signals.

The number of circuits needed in a pool is calculated from a forecast of the number of call attempts requiring that particular type of circuit during the busy hour, on the basis that each call attempt will engage one circuit for a very short but predictable time (which varies only in respect of the amount of routing information involved for each particular call). The proposed circuit numbers cannot easily be verified by a prospective private-system purchaser, because both the specification of performance standards and the calculation of circuit quantities require a detailed knowledge of the operation of the particular switching system involved. Underprovision will quickly show up as an excessive number of failures of a particular type of call attempt, and the potential cost of this should be weighed against the cost of providing a more generous pool of circuits. It is a wise precaution at least to seek an assurance from the supplier that the standards being implemented are appropriate to the needs of the business. The public-network purchaser is better placed in this regard, having the technical backup needed to ensure that appropriate performance standards are specified and applied.

The preceding paragraphs have drawn attention to areas of telephone switching-system design where it is particularly important that capacity considerations are properly taken into account when a system is being dimensioned for a particular switching-centre or PBX installation. Another equally important area of concern is the *call-control processor*, which may also limit system capacity and is dealt with in Section 7.4. Traffic-routing considerations can affect system complexity and cost, and are discussed in Section 7.5. Clearly, the quoted initial cost of a new switching system can be misleading, since the cheapest system that appears to meet the purchaser's needs may not have been dimensioned to the same standards as a more expensive alternative, which may prove to be a better long-term investment. Future growth requirements must also be taken into account, in terms of (a) the ultimate capacity likely to be needed within the expected lifespan of the new switching system and (b) the cost and probable timing of intermediate growth steps towards that ultimate capacity. The estimation of this long-term cost may be complicated by the fact that, although the capacity of a well balanced switching-system design should not be limited by one element more than any other, it is usually the case that the growth steps for different elements do not coincide. Ideally, a comparison using *discounted cash flow* (DCF: see Section 11.3) should be made of the long-term costs of all available switching systems that are technically suitable for the particular communications need.

A design team that gets it wrong too often, whether independent or employed by a particular manufacturer, is unlikely to survive in this very competitive environment. Nevertheless, a few questions asked on the basis of even limited knowledge can be helpful in ensuring that a new switching system will perform satisfactorily at the quoted cost. An established switching centre or PBX can also be a source of problems, even when a soundly designed switching system has been correctly dimensioned to the best possible forecasts, because unforeseen factors may cause the level and distribution of traffic to be very different from that predicted. If particular types of call attempt fail too often or excessive congestion is encountered, it is likely that the causes will be those discussed in relation to a new switching system. The pointers that have been given to typical design problems can, therefore, be equally applicable where consideration is being given to the extension of a switching centre to meet changed demand. It is a worthwhile investment to periodically engage a consultancy organisation to carry out a detailed traffic survey; this involves supplying appropriate measuring equipment, analysing the measured data and advising on any action considered necessary.

7.3 Data switching system

Much that has been said about the general features of telephone switching-system design is also applicable to public telex switching systems, with message storage capacity an additional factor to be considered. Whereas the early electromechanical analogue systems bore a very close resemblance to the equivalent telephone systems of their day, modern digital message-switching telex systems are structured quite differently, having a much closer relationship with true data switching systems employing the packet-switching principle. Message-switching and packet-switching systems, although quite different in their detailed operation and in no way interchangeable, are affected by certain common traffic-related design considerations, as described in the following paragraphs.

A message-switching or packet-switching centre is built up on a modular basis from the elements of a specially designed switching system in much the same way as a telephone switching centre. There are fewer functional elements to consider, however, because data switching systems are not required to provide as many different permutations of facilities or cover as large a capacity range as telephone switching systems, and self-routing messages and packets make traffic-route signalling circuits unnecessary. In consequence, fewer types of equipment module are needed, in smaller quantities, and the scope for precisely tailoring a system to a particular requirement is more limited. As with telephone switching systems, each data switching system provides a unique choice of design features and facilities for switching centres of a given type and range of sizes, but the choice of system is restricted because there are fewer manufacturers in this field.

A data switching system can be broken down into functional elements in the same way as a telephone switching system (Fig. 7.2). The traffic-related design considerations for a *control processor* are broadly similar whatever the type of traffic being switched, as discussed in Section 7.4. The two other functional requirements in common with a telephone switching system are local-line termination and switching. The equivalent of a telephone *line circuit* in a data switching system is usually known by a system-specific name, for example 'terminator'. This is also required on a per-line basis, and once again there are similar arguments in respect of spare capacity (to cater for possible future needs) versus cost. The *switching element* of a packet-switching system is necessarily a digital switch, and modern message-switching systems also employ digital switching. The switches are limited in capacity, as is a telephone-system switch-block, by virtue of the number of connection points provided; hence similar capacity-limiting considerations apply. Because there is no need for signalling over the circuits of traffic routes, the routing information being carried in the messages or packets, these circuits are terminated directly on the switch.

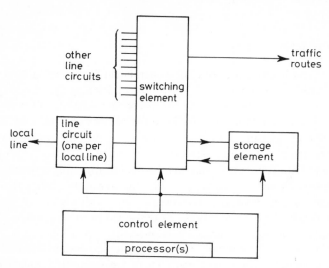

Fig. 7.2 *Principal cost-influencing functional elements of a data switching centre*

Each of these three functional elements has a parallel in a telephone switching system, but this is not true for the *storage element*, which is unique to message-switching and packet-switching data systems. In modern digital systems, it is common practice for the processing, storage and switching functions to be incorporated in a single control, storage and switching *equipment module*, usually known as a *subsystem*. Within a subsystem, the capacities of the separate functions are related on a basis determined by the system designer from typical network characteristics; they are (a) the message or packet handling throughput capacity of the control processor, (b) the capacity for storing messages or packets, and (c) the maximum number of line terminations and traffic-route terminations catered for by the switch. Each subsystem also accommodates a number of line-termination circuit modules, which are fitted as required up to the permitted maximum. One or more of these multifunction subsystems, each of which, together with its associated line-termination circuits, is a complete switching system in its own right, will be installed to provide a new switching centre with the required overall capacity. This overall capacity is easily increased by the provision of an additional subsystem, as and when required.

The points made at the end of Section 7.2 concerning the initial and long-term costs of a switching system are equally valid here, although there is less scope for optimising choice. The implications of the relationship between switching, traffic-route and storage capacities have already been discussed in Section 6.5 and, because of the effect on processing needs and network design, the subject is further explored in Section 7.4.2 and Chapter 8. When selecting a system for a new switching centre, it is important (a) to ensure that comparable traffic-related standards have been used in the calculations for competing quotations, and (b) to make the choice by taking into account not only the initial cost but also the future costs of switching-centre extension. Where subsystem or equipment-module capacities differ between systems, as they usually do, the timing of extension costs will be unique to each system.

7.4 Processor control

The control of a telecommunications switching centre by computers, referred to in this context as processors, may be effected in different ways. The control functions may be (a) centred on a single main processing unit, (b) shared among a number of separately located microprocessors, each allocated defined tasks within the overall controlling process, or (c) mostly centred on a main processor with subsidiary tasks devolved to microprocessors. Methods (a) and (b) are known respectively as *centralised control* and decentralised or *distributed control*; method (c) is a mixture of the two. Centralised control is the most common, but large public-network systems often employ a mixture of centralised and distributed control.

Conventional mainframe or desktop computers have proved to be adequately reliable for their purpose, but they are not required to give the uninterrupted service over many years demanded of a processor in a telecommunications switching centre. Switching-centre processors, therefore, are usually specially designed or adapted for the purpose. Improved reliability is achieved by the duplication or triplication of vital components and, for all but the smallest PBXs, the processors themselves are similarly replicated in some way.

The task of controlling any telecommunications switching centre increases in complexity with size and the range of facilities offered. Control is actually exercised by computer software in the form of one or more programs running on a processor, not by the processor itself: hence the term *stored-program control*. The controlling program(s) and all other information relative to the operation of the switching centre, including traffic records, are stored as computer data, together with user information such as call records, entitlement to optional facilities, etc. The stored data can be changed by inputting new information from an associated data terminal or loading a replacement program, thus eliminating the need to alter wiring or replace equipment.

7.4.1. *Telephone switching-system control*

The designer of a telephone switching system is faced with a basic choice between centralised or distributed control; the decision, once made, is irreversible since it influences every aspect of system design. A centrally controlled switching system equipped with a single microprocessor offers the simplest solution to the designer. This concept is, however, restricted to smaller-capacity systems where processing demands are limited, particularly PBXs. One controlling program only is needed, and it is possible to provide an adequate reserve of processing power to ensure that the processor cannot be overloaded. The larger the switching system, the more complex is the problem faced by the system designer. The task of controlling a large telephone switching centre is so complex that it has to be split between a number of separate programs, each of which is called upon for a particular stage or aspect of the overall task. The computing demands for such a system are too great for a single centralised processor and, in practice, a number of separate processors are used, operating as an integrated multiprocessor unit; the modular construction allows a switching-centre designer to match processing capacity to the demands of a particular switching centre. Whenever control is either (a) centralised in a large main multiprocessor, with or without subsidiary microprocessors, or (b) fully distributed among many microprocessors, the various processors must communicate with each other and with the information storage areas for the switching system to operate successfully. Intercommunication is effected by the exchanging of data messages, requiring the processors and storage areas, including security duplicates, to be linked by what is, in effect, a data network. This can give rise to traffic-flow problems; competing messages could simultaneously attempt to access the same point in the network and a situation could arise where the maximum rate at which a processor can deal with messages is exceeded. In consequence, storage space has to be provided for messages to be queued and they can, therefore, be subject to delay. It is more difficult to achieve efficient intercommunication between processors with fully distributed control, but this does offer some potential advantages over centralised control in terms of flexibility and system reliability, because the processing power is spread over more but smaller units which can be co-located with the equipment they control.

Processing effort, including the exchanging of messages between processors in a multiprocessor system, is only needed (a) to set up a call, (b) to a lesser extent to clear down a call, and (c) in some modern systems to change conditions on a user's instruction whilst a call is in progress, as happens when a supplementary service is invoked. An increased number of call attempts, therefore, means more work for the controlling processor(s). Every processor has a limited workload capacity, specified in terms of *busy-hour call attempts* (BHCA); hence the internal data network of a multiprocessor switching system can become overloaded and unable to cope with the flow of control messages if the call-attempt rate significantly exceeds the designed switching-centre capacity. The effect would be exactly the same as if, under normal load conditions, one or more of the processors failed owing to a fault, thus reducing the available processing capacity. In an extreme situation, the collapse of intercommunication between processors would quickly result in loss of control and possibly the indiscriminate cutting off of calls in progress. The switching centre would temporarily be out of action, and restarting the processors would prove difficult while the rush of call attempts – now increased by users trying to re-establish lost calls – continued.

The loss of calls in progress must, therefore, be avoided if at all possible. This means that the load on the processors of a multiprocessor switching system must be controlled. Typically, the various types of processing task are graded in priority order so that those of least importance can be automatically deferred or rejected when demand is heavy. This controlling function is generally known as *processor load control* and is required to ensure that, when a switching centre is subjected to an abnormal rush of call attempts or a fault reduces the available processing power, sufficient processing power is reserved to complete and clear any calls already in hand. This is achieved by preventing the processors from (a) dealing with non-urgent 'management' tasks and (b) accepting new call attempts. Processor load control must also restore normal service as soon as the overloading problem is resolved. Some call facilities and calls over particular types of circuit use more processing power than others; hence a count of the call-attempt rate is inappropriate and the actual processing workrate has to be monitored. The most sophisticated systems reject new call attempts in a progressive manner; some processing power is initially reserved to handle emergency calls and return busy tone to non-priority callers, and only in extreme circumstances are all new call attempts rejected indiscriminately without tone. Some systems, however, reject all call attempts immediately and give no indication to callers that the switching centre is at fault; this means less system complexity and hence reduced cost. Overloading problems of this nature should, of course, occur only very rarely if a switching system is reliable and adequately dimensioned for its task, and the additional cost of a more efficient processor load control must be weighed against the possible costs of occasional user frustration.

It will be apparent from the foregoing that, particularly for large switching systems, the design of the control area and its processors can have a major influence on the standard of service offered by a telephone switching centre. Efficient processor load control is especially important where there are several interconnected switching centres in a network. It is then possible for uncontrolled overloading at one switching centre, resulting in the loss of calls in progress and the inability to accept new call attempts, to spread through the network. Some of the failed connections and new call attempts will involve the interconnected switching centres and, because the failed centre is unable to accept new call attempts directed to it, these other switching centres may also become overloaded and fail in turn. It is inconceivable that the whole of a large public telephone network could be put out of action in this way, but a complete network failure is a distinct possibility in the case of a smaller private network. The ability of a system to cope with overloading is reflected in its cost, and it is important that switching-system purchasers should know exactly what level of security is offered by the systems under consideration. The right questions, put to potential suppliers at an early stage in purchasing negotiations, can draw attention to the merits and demerits of different systems in a way that would not otherwise be possible. The purchaser can then be confident that the chosen system will offer a level of security consistent with the needs of the network in which it will be required to operate.

7.4.2 Data switching-system control

Data switching systems are both simpler and smaller than the very complex public telephone network systems, and multifunction subsystems as described in Section 7.3 make a processor per subsystem the ideal design choice. This is a form of centralised control because each subsystem is controlled independently. Processor intercommunication is required only between subsystems, usually over a multipurpose data highway. This highway interconnects all subsystems and its primary task is to transfer user-originated messages or packets which are being switched from one subsystem to another. It is the throughput capacity of this highway that limits the number of subsystems which can be linked together as a single switching centre. The store-and-forward capability of message- and packet-switching systems removes the need for a complex processor load-control scheme of the type described for telephone systems. Traffic arriving at a switching centre is automatically queued in the incoming storage area before control is passed to the associated processor for onward switching, at a rate which that processor is able to accept. The fixed storage capacity automatically imposes an upper limit on the amount of traffic that can be accepted by the switching centre; any traffic in excess of this limit is rejected and has to be retransmitted over the link (from user's terminal or previous switching centre), to be treated by the receiving switching centre as fresh traffic. Provided the highway capacity is adequate for the maximum possible inter-subsystem demand imposed when accepted traffic is at the highest possible level, the task of processor load control is restricted to reducing the possibility of within-subsystem overloading as a result of a processor fault.

The subsystem approach to switching-system design places a responsibility on the system designer to achieve a satisfactory inherent balance between storage and switching capacities, together with ample processing power. A correct balance is particularly important for a packet-switched network, because of the higher speed of operation and the generally tighter capacity tolerances. The network designer, on the other hand, is given little scope to influence the situation. Assuming that a switching system is suited to the job required of it, the network designer does not really *need* to do more than equip the linking traffic routes to an appropriate occupancy for the expected data flows (see Section 8.2). However, if the level of traffic to be switched at peak periods proves to be in excess of that forecast, both network delay and the amount of traffic rejected may prove to be greater than considered tolerable. A switching system that is not suited to a particular network environment, especially where its capacity balance is different from those of other switching centres in the network, could then increase these problems (this is why packet-switched networks are generally equipped with the same type of switching centre throughout, and is one of the reasons why switching-centre growth is planned on a network-wide basis).

In a data network, whether message switched or packet switched, the occurrence of a switching-centre fault which reduces processing capacity during a critical peak traffic period could, without effective processor load control, result in the failure of that switching centre. Under extreme conditions, this problem could spread to other switching centres as in a telephone network, and a total public packet-switched data-network failure of this nature has occurred outside the United Kingdom. The potential purchaser of a data switching system should, therefore, be aware of the possible consequences of operating a system without adequate safeguards against processor overloading.

7.5 Switching systems and traffic routing

Where a number of switching centres is interconnected by a mesh-type network, as in all public telecommunications networks and some large private telephone and data networks, it is possible to route traffic to a particular destination in a variety of ways. Routing is controlled by the switching centres, which operate to a predetermined set of rules designed to enable them to select the required routing for any chosen destination. This function places a demand upon switching-system processing capacity and also influences the cost of each particular switching centre through its effect on traffic-route capacity needs. The advantages and disadvantages of different methods of routing traffic are discussed in Chapter 8 but, because switching-centre processing demand and dimensioning are affected by the routing method employed, it is also necessary to outline some of the available routing options in the following paragraphs.

Traffic outgoing from a switching centre and destined for a particular distant switching centre is either carried over a direct link between the two centres or, where no such link is provided, switched indirectly through one or more intermediate *tandem switching centres* (Section 8.1). *Direct routing* is universally possible only in a so-called *fully connected* network. Most networks are *partially connected*, which means that only the more heavily used routes are direct and the remaining connections are effected over tandem-switched *indirect routings*. Indirectly routed traffic is first routed to a switching centre which is (a) accessible over a direct link and (b) suitably located. It is then switched to another direct link, either to the destination switching centre or, if this is not possible, via another tandem switching centre, and so on. Generally the shortest possible routing is used, ·but this may still require more than two stages for some connections.

Although several different routings may be possible, the simplest method of directing traffic from one switching centre to another is to use only a single, fixed

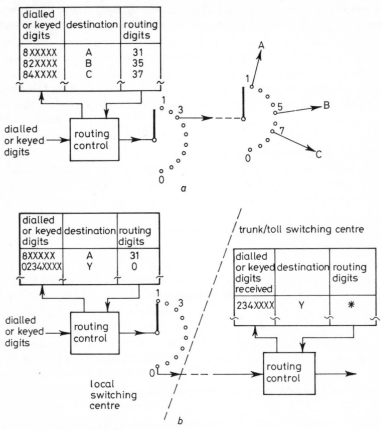

Fig. 7.3 *Single-choice routing:*
(a) Principle of routing control
*(b) Transfer of routing control for trunk/toll calls. The symbol * indicates digits necessary to route call to destination local switching centre via other trunk/toll switching centres as required*

routing; this method is henceforth referred to as *single-choice routing* to distinguish it from more complex routing methods. It was for many years the only routing method available because of the limitations imposed by electromechanical switching systems, yet it is an option which, because of its simplicity, continues to be appropriate for many telecommunications networks. The first switching centre to which a user is connected normally controls the routing, by making use of what is in effect a conversion table. This table details the routing for every possible destination in the network and, because different routings are required from each switching centre, it is unique to one centre. In electromechanical switching systems the table is in the form of wired circuitry, but modern stored-program-control electronic systems use a look-up table that is part of the processing software. When single-choice routing is applied to a telephone-network switching centre, for example, the dialled or keyed code digits received from a caller are translated by reference to the conversion table into the appropriate routing digits needed to steer the call set-up through the network to the required destination (Fig. 7.3a). In a public network, the local switching centre controls the routing of local calls only, handing over control to a trunk or toll switching centre on receipt of the trunk or toll prefix digit(s) (see Fig. 7.3b); the trunk or toll switching centre in turn passes control to an international switching centre when the prefix digits indicate this need.

Alternative routing, often known in the United States as alternate routing, makes it possible for a switching centre automatically to select from a number of different routings, as explained more fully in Section 8.1. This method of routing traffic only enables traffic congestion to be bypassed but also offers network cost savings. The need for the switching-centre control element to make a choice each time a connection is required complicates the route selection process, however, and alternative routing schemes have only become widely available as a practical network option with the introduction of electronic switching systems and, in particular, stored-program control. Since alternative routing can be applied at every switching centre involved in a particular connection, all share in the control of routing as call set-up proceeds. The choice of one routing in preference to another may be made either (a) on a fixed basis, where a circuit on a single nominated second-choice routing is selected when all circuits on the normal first-choice routing are in use (Fig. 7.4a), or (b) on a dynamic basis, where the first available circuit from a number of possible routings is selected if no direct circuit is available (Fig. 7.4b).

Fixed alternative routing can be expected to use marginally more processing power than single-choice routing because the conversion table at each switching centre is effectively given another column of routings, referred to in the event of there being no free circuit capacity on the first-choice route indicated by the first column. *Dynamic alternative routing* offers a greater choice of possible routings at each switching centre and also marginally adds to the demand for processing power. Whichever form of alternative routing is used, the additional processing power will be required at peak traffic periods when overall demand is at a high level; it is, therefore, particularly important that a switching system has efficient processor load control. If the full economic benefits of alternative routing are to be realised, it is necessary to apply grades of service (GOS) which differ from those used to dimension the routes of a single-choice-routed network, as discussed in Chapter 8. Since GOS determines the number of traffic-route terminations needed at each switching centre, this influences switching-centre cost. Switching-centre cost is not by itself a guide to the economics of introducing alternative routing, however; it is necessary to compare overall network costs using different routing methods.

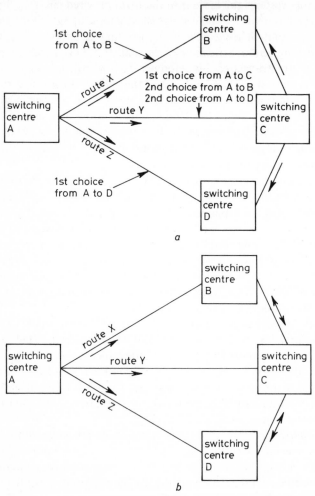

a

b

Fig. 7.4   *Alternative routing. Routings shown from switching centre A only*
(a) Fixed: second-choice circuit selected only when all first-choice circuits are busy
*(b) Dynamic: first available circuit on any other route selected if no direct circuit is
available. Process is then repeated as necessary at intermediate switching centres*

Modern telephone switching systems are increasingly capable of operation in any one of the three basic routing modes, but there are still many switching systems which offer restricted options in this respect. The desired facility must be either already built into the controlling software or available as a standard additional or replacement package. The fact that circuit switching is used allows different traffic-routing options to be applied to parts of a network or even individual switching centres, provided the affected traffic routes are appropriately dimensioned. However, the use of alternative routing, particularly of the dynamic variety, must be closely controlled as part of an overall network policy because its effect on route traffic levels complicates the traffic forecasting process and can extend beyond the immediate area of its introduction.

Data switching systems have been in the forefront of a change to more complex routing schemes. Packet-switching systems are required to switch self-routing addressed data packets which do not need to be routed over a predetermined fixed path in order to reach a particular destination (Section 3.3); dynamic alternative routing has, therefore, always been an inherent feature of their design. Each packet is interrogated by a switching centre to determine its address. If this address is of a terminal which is not connected to that particular switching centre, the first available free capacity on an outgoing circuit is taken, whether it is on a circuit direct to the required destination switching centre or to another intermediate centre. Once an indirect routing has been selected, the same procedure may need to be followed at several intermediate switching centres in turn until the destination switching centre is eventually reached. The only routing choice available to the packet-switching network operator is that between datagram and virtual-call working (Section 3.3); since both affect the operation of the network as a whole, the decision between them must be made on a network-wide basis.

Data messages are also self-routing in a message-switched store-and-forward network, but in this instance alternative routing is an option and not an essential feature of the mode of operation of the network. Single-choice routing or alternative routing of either the fixed or dynamic variety can be used as in a telephone network and the same considerations affect switching-system design. It is similarly unnecessary to restrict choice to one routing method network-wide; each switching centre can be considered individually, provided (a) the affected traffic routes are appropriately dimensioned and (b) the introduction of alternative routing is closely controlled on a network-wide basis.

7.6 The switching of integrated speech and data traffic

Some types of data traffic have been handled by PBXs and public telephone networks for many years, within the limitations imposed by (a) circuit switching, because of the time needed to set up calls, and (b) analogue transmission techniques, which restrict transmission speed (Section 3.1.1). Public packet-switched networks meet the demand for transporting data traffic unsuited to the telephone network, but separate digital access has been necessary to enable users to take advantage of higher transmission speeds. Although there is as yet no alternative to the retention of circuit switching for telephone networks, this situation is now changing. Public telephone networks are evolving towards the integrated-services digital network (ISDN) concept (Section 4.1), through the introduction of stored-program-control digital switching systems and the elimination of analogue transmission. Digital fourth-generation integrated-services PBXs (ISPBXs) (Section 5.4), which incorporate both circuit and packet switching, are being introduced for private networks. These developments enable the complete integration of speech and data traffic over ISPBX and ISDN digital local circuits, removing the restriction on data transmission speed formerly imposed by analogue transmission; this permits common digital access to both the ISDN and the existing packet-switched network.

The design of public-network switching systems is not significantly affected by these developments because the method of switching is unchanged. The forecasting process must, however, take account of any changes in average traffic characteristics as a result of an increased proportion of data traffic. By contrast, the ISPBX is a new concept which is at present required to separate packet-type data from other traffic so that this can be switched independently. Data traffic is likely to make up a greater proportion of the total traffic load than with PBXs of an earlier generation. A general system-design pattern has not so far emerged and there is no alternative but to check and compare the claimed merits of competing systems, preferably on the basis of independent technical advice.

The limitations of the public ISDN stem from its continued use of circuit switching, which is unsuited to the transmission of some types of data traffic (although access to an existing, separate public packet-switched data network is available over ISDN local circuits to circumvent this problem). The all-purpose public telecommunications network of the future is expected to use an enhanced form of packet switching known as *variable-bit-rate switching* (Section 4.2), with speech, data and video signals all being transmitted as interleaved data packets over the same channel. This concept also offers the most promising solution for the next generation of PBXs and LANs (Section 5.4), where it is likely to find application long before it can be brought into widespread use in public networks. The combination of variable-bit-rate switching with the high operating speed and enormous capacity offered by optoelectronic technology can be expected to make possible the design of switching systems that are free from many of the inherent traffic-related restrictions of current systems.

Chapter 8

Traffic considerations in network planning

8.1 Telephone networks

The basic components of a telephone network have already been identified; telecommunications users are connected by *local circuits* to a *switching centre* and, where the users are sufficiently numerous and widely located, there are a number of switching centres interconnected by *traffic routes*. Earlier chapters have also drawn attention to various traffic-related factors which affect network performance. These and other factors now need to be considered from the point of view of their collective influence on the planning of a telephone network.

The planning of a *public* telephone network is traditionally approached from what might be termed the bottom up, the initial planning objectives being (a) to divide the overall area to be served into local switching centre catchment areas of the requisite size and (b) to locate each local switching centre within its catchment area so as to minimise overall costs. Local circuits are normally provided on a per-user basis; a forecast of traffic flow is required only where, for a link between a PBX and the public network, the amount of traffic justifies more than one public network local circuit. The length of a local circuit is limited by the type of cable employed and the inherent design of the local switching-system terminating circuitry. Public-network local circuits are normally no more than a few kilometres in length. However, longer non-standard circuits are used where a telephone service cannot otherwise be provided; exceptionally, for business reasons, a user may need to have a telephone number on a distant local switching centre, in which case an amplified out-of-area local circuit is provided at extra cost. It follows that the catchment area of a public-network local switching centre is usually of about the same physical size whether it is connected to many thousands of users or a few hundred. Unusually small catchment areas are found where the number of users within a normal-sized area would exceed the maximum capacity of a single switching centre – a situation typical of business districts in large cities. At the other extreme, a catchment area in a remote rural district may need to be larger than normal to cater for small numbers of outlying users whose location precludes service by other than non-standard means. In general, community of interest and the disposition of the potential users determine the number and siting of local switching centres, and user density is typically plotted on a map of the area to be served so that catchment areas of an appropriate size can be centred where the users are most concentrated.

The largest contribution to local telephone service cost is that of the cable network; hence the most important criterion for local switching-centre location within a catchment area is *minimum cabling cost*. Computer programs are commonly used to determine the theoretical central point of the catchment area, calculated on the simplistic assumption that each local circuit will be cabled in a straight line between user and switching centre. Practicable cabling routes are then considered, the nearest intersection of main cable routes to this point being identified as the preferred location for the switching centre. It may, of course, prove impossible to secure a switching-centre site of the requisite size in the locality, but the further away a switching centre is from the optimum location, the greater will be the cost of cabling. The differences in cost for a large switching centre can amount to huge sums over a few hundred metres, and cabling costs are, therefore, an important factor to be considered when deciding between alternative sites.

Switching-centre catchment areas for *private* telephone networks are often confined to one site or a group of nearby sites where any limitation on local-circuit length is unlikely to be a problem. However, where outstationed extensions must be served, use often has to be made of leased private circuits over a public network for this purpose. The location and number of PBXs required should ideally be determined by balancing PBX costs against the cost of both the internal cabling and any private circuits required for tie lines and/or extensions, within the limited range of options allowed by the relative disposition of the users to be served. A private network large enough to need several PBXs may be required to serve hundreds of users concentrated at a few large sites or, at the other extreme, an equally large number of users dispersed among many small sites. Concentrated networks are relatively straightforward to design since each site is usually served by a separate PBX with common catchment-area/site boundaries; only where two or more sites are relatively close together is it likely to be both practicable and economic to serve them by one PBX. A widely dispersed network presents a more complex and unique problem to the network designer and will have a greater need for private circuits to interconnect the various sites. Where alternative locations for a PBX are available – either within one extensive site or at different sites – and there are no other overriding factors, minimum cabling cost is again an important factor in the overall cost equation as for a public network.

The full procedure for local switching-centre or PBX location is only applicable in a rarely enountered green-fields situation. Significant changes to the boundaries of a local switching-centre catchment area are rarely needed in an established public telephone network, although site acquisitions or disposals commonly alter private-network **PBX** requirements. Where boundary changes need to be made, the existing network of cables will heavily bias the optimum location to the present switching-centre or **PBX** site. However, there is the occasional need to replace an existing local switching centre or **PBX** because of excessive maintenance costs, the inability to obtain spare parts or the need to introduce new facilities. Where, for example, user growth is concentrated in one part of the catchment area and alternative sites at which to install the replacement equipment are available, a reassessment of the optimum location may then be justified.

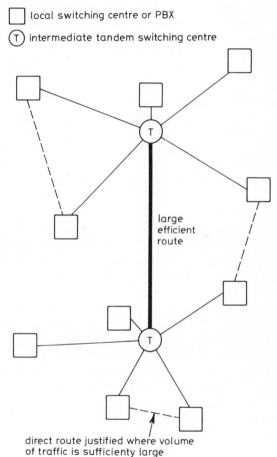

Fig. 8.1   *Partially connected network.*

Having determined the number, location and size of local switching centres or PBXs, the designer must next decide how they should be interconnected. Local switching centres and interconnected PBXs are directly associated with the users they serve, forming the periphery of a network. Each must be structured internally to optimise its traffic-carrying capacity, but traffic considerations have no other influence on the planning of this part of the network. They are, however, crucial to the design of the interconnection arrangements – the central core of the network – where costs can only be minimised by achieving the best possible traffic-carrying efficiency. A *fully connected network* in which all switching centres are directly connected to each other is unlikely to be economically viable unless the majority of routes is adequately loaded, a condition that is more difficult to achieve as network size increases. The actual break-even point for each particular network depends on the traffic distribution and switching/transmission cost relationship. As the number of local switching centres or PBXs in a network grows, the traffic flow between them will also increase – to the point where, as the balance between switching and circuit costs changes, it becomes economic to interpose an additional tier of intermediate tandem switching centre(s) between them, creating a *partially connected network*. Inter-switching-centre traffic is then concentrated on to a few large and, therefore more efficient traffic routes (Fig. 8.1), so that the combined cost of these normally shorter routes and the additional tandem switching centres is less than that of the fully connected option of many very long, small, inefficient routes. In a public network, tandem switching always proves to be cost-effective. This is not necessarily the case for the smaller private network, however, since route sizes are often too small for any gain in efficiency from an increase in size to influence the cost balance.

In a partially connected network, a comparison of the direct circuit cost with the cost of indirect tandem circuits plus switching decides whether a direct route can be justified; if not, the traffic is routed indirectly. Some of the routes, therefore, will be carrying not only directly routed traffic between the terminating switching centres but also indirectly routed traffic between these and other switching centres. This requires traffic forecasts originated on an inter-switching-centre basis to be summated to provide forecasts on a traffic-route basis (Fig. 8.2).

Each intermediate switching centre in a public telephone network effects long-distance interconnections for a group of local switching centres. This allows the local and trunk/toll calling functions to be separated. In very large public networks it can prove economic to provide another, second tier of *trunk/toll switching centres*. Where traffic is progressed in an organised manner through two or more distinct tiers of switching centres in this way, the network is termed *hierarchical*. The problem is, however, to decide just how many intermediate switching centres should be provided. This was, in the past, often determined on purely practical grounds; in established networks, for example, trunk/toll switching centres were sometimes located as direct one-for-one replacements of the original trunk/toll manual switchboards in order to take advantage of existing cable routes. However, in a network with a given amount of traffic to be switched, a reduced number of larger trunk/toll switching centres means fewer, larger traffic routes, all contributing to cheaper per-circuit switching and transmission costs. The cost reductions with increased size differ between switching and transmission systems, and each has its upper size limit; hence there will be an optimum number of switching centres for each network which minimises overall network costs.

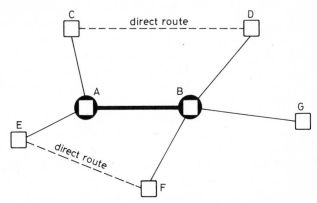

Fig. 8.2   *Summation of inter-switching-centre traffic forecasts to provide a traffic-route forecast. Inter-switching-centre bothway traffic flows which use route A–B:*

Traffic flow	erlangs	CCS
A–B	70	2520
A–D	15	540
A–G	10	360
A–F	12	432
C–B	16	576
E–B	20	720
C–G	8	288
C–F	4	144
E–D	6	216
E–G	5	180
Total flow A–B	166	5976

The complete replacement of all analogue switching and transmission equipment by digital plant provides the opportunity for an established public network to be restructured to an extent that has never before been possible, enabling a cost-saving reduction to be made in the number of trunk/toll switching centres. The optimum number of trunk/toll switching centres usually proves to be considerably fewer than already exist; for example, about 60 instead of almost 400 in the British Telecom network. Reducing the number of switching centres, however, means fewer routing alternatives, larger switching centres and routes, and a greater chance of excessive congestion if a switching centre or route fails. This problem is compounded by the changed equipment modularity which results from the replacement of analogue by digital transmission systems; the use of 24- or 30-channel modules instead of 12-channel makes it more expensive to provide small direct routes, to the extent that some existing routes are no longer commercially justifiable. It may be necessary to retain more than the optimum number of routes in order to maintain a large enough number of possible routings, as a safeguard against switching centre or route failure. Questions such as how many trunk/toll switching centres are appropriate for a large public telephone network should, therefore, be considered in relation to network structure and traffic routing in the widest possible sense. For a smaller private network, structure and traffic routing can usually be considered independently, the number and location of any tandem switching centres being decided on the basis of a straightforward cost comparison and practical considerations such as site availability.

When the requisite number of public-network trunk/toll switching centres has been decided upon, catchment areas are formed from groups of local switching-centre catchment areas, taking into account community of interest among users and topographical features. The local switching centres within each trunk/toll switching-centre catchment area are the 'users' in this situation. The optimum location for minimum cabling cost will often prove to be coincident with or close to that for a local switching centre; hence it is a common practice to house both trunk/toll and local switching centres in the same building.

The bottom-up procedure of network planning so far described has evolved in a monopoly public-network environment, and is not necessarily appropriate where a new competitive network is being established alongside one that already exists. The Mercury Communications network in the United Kingdom, for example, was established on the top-down principle, with a trunk network as its first objective. This has enabled bulk business traffic to be carried from the outset, with the expansion of the user base being followed up as and where conditions permit.

Single-choice routing is commonly used in private networks, and is still widespread in large established public telephone networks because of the restrictions imposed by the residual electro-mechanical switching centres, a large number of which have yet to be replaced. As these public networks are re-equipped with digital switching centres and transmission systems, however, the simple choice between direct and indirect single-choice routing is widened to include the *alternative routing* methods introduced in Section 7.5. Advantage can also be taken of the wider choice of routing options offered by modern PBXs for private networks, including the automatic selection of the least-cost route to any particular destination where competitive public-network carriers exist, and a form of fixed alternative routing which causes excess traffic to overflow from a private circuit to the public network. Alternative routing, in its various forms, offers scope for increased traffic-carrying efficiency and consequent reductions in overall network cost, and should be considered as a possible option for any tandem-switched telephone network that can be suitably equipped, whether public or private.

Fixed alternative routing (see Section 7.5 and Fig. 7.4a) provides access to a second-choice route in the event of there being no free circuits on the first-choice route to the required destination switching centre. This approach allows callers a second chance to establish a connection at each switching stage where it is applied and congestion is encountered, thereby increasing the probability of a successful call set-up. Both first- and second-choice routes are predetermined and invariable at each stage.

In a tandem-switched network, the routes to which fixed alternative routing is to be applied are divided into two categories; these are (a) first-choice-only routes, and (b) first-choice routes which are also nominated as second choice, taking overflow traffic from the former and each other. The designated first-choice-only routes are dimensioned to a poorer grade of service (GOS) than would be the case with single-choice routing. This has the effect of providing fewer circuits for a given amount of traffic, which forces some calls to overflow to the nominated second-choice route at busy periods. Overflow commences before traffic flow reaches the forecast busy-hour level, at a level of traffic flow that would be within the capacity of a normally dimensioned route. This ensures maximum utilisation of the route over a longer period, giving rise to the term *high-usage route*. These routes are, therefore, *more efficiently* used than routes dimensioned to normal single-choice routing standards. Overflow traffic from these first-choice-only routes is concentrated on to the remaining routes, selected to take second-choice traffic in addition to the first-choice traffic for which they are primarily intended. There is a danger that, at busy periods, overflow traffic could monopolise such a route to the detriment of first-choice traffic; this problem is overcome by barring a few circuits on each route from the overflow process – in effect, reserving them for the exclusive use of first-choice traffic. In addition, these routes are *fully provided* (Section 6.5) to carry the forecast busy-hour first-choice plus second-choice traffic; they are, therefore, *less efficiently* used than routes dimensioned to normal single-choice routing standards. By minimising the number of these less efficient mixed-traffic routes and maximising the number of high-usage routes, fewer circuits are required than would be the case with single-choice routing and overall network cost savings of about 5% are possible. With this dimensioning approach there is also a clear distinction between the majority first-choice-only routes and the few mixed-traffic routes, only the latter having to be dimensioned on the basis of aggregated first-choice and overflow traffic. This relative simplicity is particularly advantageous where fixed alternative routing has to be implemented on a piecemeal basis.

In one sense, private networks offer the network designer a wider traffic-routing choice than public networks, because the traffic to selected destinations can be routed over a public switched network instead of being confined to inter-switching-centre circuits that are exclusive to the private network. The previously mentioned form of fixed alternative routing, which allows excess traffic to be diverted from a private traffic route to a public network, further widens the scope for optimising costs. Essentially, the objective of using this form of alternative routing is to minimise leasing charges by reducing the number of private circuits required, effectively making them high usage as in a public network. The reduction in the number of circuits ensures that each circuit is loaded to a high occupancy throughout as much of the working day as possible, thus enforcing the overflow to the public network of any traffic peaks. It follows that the number of calls overflowing to the public network will be at its maximum value during the busy hour, when call charges are likely to be at their highest level. If too much traffic overflows, the additional call charges will outweigh any savings made in leasing costs. Depending on the traffic profile through a working day, the traffic level and distance involved, and the relationship between public-network call charges and circuit-leasing charges, the cheapest option may involve (a) routing all traffic via leased circuits without the facility for overflow, (b) routing the traffic via fewer high-usage leased circuits with overflow to a public network, or (c) routing all traffic via a public network. Moreover, the optimum leased-circuit loading for (b) may itself vary from route to route in the same private network. The end result may well be a network in which the routing method used differs from route to route. There is, therefore, no alternative but to determine the cheapest option for each route *separately*.

Two possible methods for finding the optimum number of high-usage leased circuits are either (a) to test the effect of applying a range of grades of service less stringent than that used for a wholly leased-circuit route, dimensioning to the conventional busy-hour traffic value (the practice followed in public networks), or (b) to test the effect of dimensioning to various traffic values less than that of the busy hour, applying the same GOS standard as that used for a wholly leased-circuit route. Either method involves a large number of calculations and the iterative process required lends itself to computerisation; it is possible to devise a fairly simple program to provide a quick and accurate solution to the complete cost comparison for a particular set of network circumstances. However, the smaller traffic levels and hence circuit numbers encountered in most private networks limit the range of practicable options, to the extent that it is possible to convert directly from 'wholly leased-circuit' numbers to 'high-usage circuit with overflow' numbers with a reasonable expectation of near-optimal results. A conversion table at Appendix B, based on typical criteria, covers the requirement for up to 25 circuits; beyond this, either of the two theoretical approaches is likely to provide a more accurate answer. Since the table is based on method (b), the same GOS standard can be applied to both wholly leased-circuit and high-usage leased-circuit routes; this facilitates the application of an overall GOS performance target to the network as described in Section 10.2.

Whatever the method used, an important point to be considered is that public-network traffic is charged for on a per-successful-call basis whereas leased circuits are charged for on a per-circuit basis, the number depending on traffic flow. Bearing in mind that traffic flow over a leased circuit includes both successful calls and call attempts that are unsuccessful (because the destination telephone is engaged or fails to answer, or there is distant-end congestion), the relationship between busy-hour traffic flow and an annual total of successful calls must be determined. Given the number of working days per year, the following information is necessary for this to be possible:-

(a) Average call duration: typically about 150–200 seconds for a successful call and 40–50 seconds for an unsuccessful call attempt.
(b) The percentage of busy-hour call attempts that are successful: typically about 65–70%.
(c) The relationship between total calls/traffic during a typical working day and busy-hour calls/traffic: typically the total is about 4–5 times the busy-hour figure.

In addition, for high-usage/overflow working, it is necessary to know the timing of traffic peaks outside the busy hour in relation to both the busy hour itself and to public-network call-charge rates (depending on circuit occupancy, 60–75% of the total overflow traffic can be expected during the morning, most of this in the busy hour). The typical values given should be used with caution, however, since actual figures can differ widely from business to business; every effort should be made to determine values that are *specific* to the network under consideration.

Dynamic alternative routing (see Section 7.5 and Fig. 7.4b) involves the selection of the first available free circuit on any possible routing, detected in either a cyclic or a random manner, where none is available on a direct route. The wider choice of routings made available enables consistently shorter set-up times to be obtained across a busy network than with any other method of routing traffic, but careful attention to route planning, forecasting and dimensioning is essential to achieve these and the network cost savings that can theoretically be made. These economies are possible because of two characteristics of telephone traffic: (a) morning, afternoon and evening traffic patterns vary, causing the busy hour of different traffic routes to occur at different times, and (b) the traffic peaks encountered in practice vary from the forecast on which circuit provisioning is based in respect of both volume and timing. The practice of dimensioning traffic routes to meet a fixed maximum-demand forecast makes it inevitable that, especially with fixed routing methods, many will be underutilised for much of the day. Dynamic circuit selection allows circuit capacity on routes which are underutilised during one period of the day to be used to carry traffic overflowing from other routes that are in greater demand at that moment, and vice versa at other times of the day. It is then possible to use the available circuit capacity much more effectively so that fewer circuits overall are needed to carry the same amount of traffic. Savings are variously claimed at up to 12% of overall network costs compared with single-choice routing.

Given the random nature of traffic and the fact that circuit selection takes place independently and on a random basis at each switching stage, there is no way of predicting which particular routing will be followed during a dynamically routed call attempt. It is, in fact, necessary to introduce controls into the routing process to ensure that a call set-up attempt does not 'chase its tail', circulating within the network indefinitely without finding an outlet to the required destination switching centre. Dynamic routing can be applied to a conventionally forecast network dimensioned to the grade of service (GOS) appropriate for single-choice routing; this may well prove to be the safest initial approach for an established network, allowing traffic measurements taken subsequently to be used for guidance in the gradual process of optimisation. Nevertheless, there is much to be gained from achieving network cost savings more quickly by using a different method of forecasting traffic and dimensioning. Any such method depends on first identifying the different traffic and busy-hour patterns involved. Dynamic alternative routing is particularly advantageous for a public trunk/toll network; a practical approach here is to consider the switching centres one at a time and, knowing the amount of traffic generated by the local switching centres each serves, to determine its total outgoing traffic level. This traffic can then be apportioned between the available routes on the basis of mathematical probability. When this has been done for all the intermediate switching centres, a matrix of probable traffic flows throughout the network can be constructed, allowing the traffic routes to be dimensioned accordingly. The degree of optimisation achievable depends largely on the mathematical apportionment method used and, since dynamic routing is still a relatively new concept, many of the ideas being generated for this worldwide are as yet untried in practice.

The reliability of switching-centre and, to a greater extent, route forecasts must to some extent be in doubt when any new method of apportioning traffic for dynamic routing is used for the first time. It is possible to provide additional circuits as a safety measure, by applying a better GOS and thus causing each route to be operated at a reduced circuit occupancy, but this runs counter to the increased efficiency possible with dynamic routing and could easily negate any savings in network costs if taken to excess. It can, therefore, be advantageous to introduce dynamic routing to a small part of the network and measure the resulting traffic levels before applying the knowledge gained to its extended use.

The most notable example of a working dynamic routing network is in the United States, where the Bell System toll network spans the country through four different time zones. The existence of these time zones significantly increases the range of different busy-hour patterns throughout the network so that it is possible, for example, to meet the peak demand in one time zone by making use of idle routings via other time zones where the peak demand has already passed or is yet to come. This permits greater efficiency to be achieved than would be possible in a network within a single time zone.

Dynamic alternative routing, although it offers significant advantages in terms of network cost savings and its ability automatically to bypass localised congestion, cannot provide the solution to all network operating problems. The ultimate solution has, for many years, been seen to require the ability to control traffic flow over the whole or part of a network from a central point, using human or computer intelligence to maximise use of the available switching and transmission capacity. This concept, usually referred to as *network traffic management* or simply *network management*, has only become practicable with the introduction of stored-program-control switching systems, some of which now permit the way in which traffic is automatically routed to its destination to be changed by external intervention. Network management requires that state-of-the-network information on system performance, traffic congestion, etc. be relayed over data links from each switching centre to a network control centre. Engineers at the control centre can then take action, either directly or by initiating a prearranged sequence of events under computer control, to divert traffic away from particular switching centres or traffic routes should the necessity arise. This ability to remotely influence the routing of traffic enables the maximum amount of traffic flow to be maintained in the event of natural or manmade disasters or major system failures. Network control centres already exist in some public networks and can be expected to be increasingly introduced to both public networks and the larger private networks as modern digital switching systems are installed to replace obsolescent analogue plant.

8.2 Data networks (excluding local area networks)

The focal point of an *unswitched data network* is a central mainframe computer, permanently linked to data terminals via the computer ports and, where necessary, leased public-network circuits (Section 5.3). Each isolated terminal or co-located group of terminals constitutes a station, the stations being grouped on a territorial and traffic/link-capacity basis. An isolated station may need to gain access to a port via an exclusive point-to-point link, whereas the total traffic generated by several conveniently located stations may be within the time-shared capacity of a multidrop link. It is also possible to link a large, remote group of stations to the computer more efficiently by multiplexing a number of data streams – each effectively a link carrying the messages to and from several terminals on a time-sharing basis – over a high-speed transmission path. The task of the network designer is to select a mainframe computer of the requisite capacity to meet future needs, size it to match the required data-handling/response-time requirement, determine the most appropriate network configuration, choose the most suitable terminal and network-interface hardware, and decide how best to provide the interconnecting links.

Given the usual random nature of terminal usage, it is uneconomic to provide a computer and its access network with sufficient capacity to carry all traffic whenever it is ready to be transmitted. When the network is busy, therefore, a terminal may need to wait for its link to become free before transmitting a message, and there may be a further delay before the computer can process that message and respond. It is, of course, possible to exercise some manual control over the load by administratively withholding certain categories of data traffic during busy periods; this ensures that both network and computer are used with more consistent efficiency throughout the working day and also minimises the capacity need. In addition, data terminals are available with a built-in store-and-forward facility to achieve an automatic load-spreading effect as well as to buffer between different terminal and link transmission speeds. However, unless the use of such terminals is accompanied by a form of priority working, delays to the transmission of messages are indiscriminate. These terminals are also able to accept incoming messages from the network at a speed in excess of their data handling capability, which brings with it a danger that data could be lost if the amount of incoming data exceeds a terminal's storage capacity. To avoid this, the release of data awaiting transmission is automatically delayed, on a block-by-block basis as and when necessary, by *flow-control* signals returned over the network from the receiving to the transmitting point.

Assuming there is no contention for access to either the link or the computer and no manual or automatic intervention, the response time measured from the input of a message to receipt of the computer's response will represent the best possible network performance. This *best possible response time* is determined by (a) the time taken to transmit data signals over the interconnecting link and (b) the response time of the computer itself. Of these two components, *link transmission time* is normally insignificant in comparison with computer response time unless a satellite link is involved. If a terminal has to wait for its link to become free before data can be transmitted, however, a variable amount of delay will be incurred; the time taken to transfer data over that link then becomes the sum of link transmission time and delay. Since queuing processes are again involved, the computer's response time similarly depends on its throughput capacity relative to the amount of data to be processed at any given time. If the instantaneous data-throughput requirement exceeds the designed processing capacity of the computer, a further variable amount of delay will be incurred which is in this case additional to the best possible computer response time. Obviously, the user is interested only in the time between terminal input and receipt of a response which is likely to be encountered in practice. This is equivalent to the overall best possible response time plus the instantaneously variable transmission and processing delays experienced during a defined busy period, normally expressed as a network-wide *average response time*. Both transmission and processing delay components are predictable on a probability basis for a given data throughput and transmission speed.

Response time can be minimised by ensuring that the network has the maximum possible *throughput efficiency*, but can only be further reduced by increasing network and possibly computer capacity, at greater cost. Throughput efficiency is actually a measure of the amount of useful information data received relative to the total amount of data transmitted, and is also referred to as the *net data throughput*. All data transmitted over a network includes some information that is of no interest to the user: start/stop or synchronising signals, error-control and flow-control signals, etc., as well as data blocks in error that have to be retransmitted. Improving the *error rate* reduces the average duration of inter-block intervals, and hence increases the throughput efficiency. By using full-duplex instead of half-duplex circuits, the turnround interval each time the direction of transmission changes over the latter is eliminated, making a further contribution to improved throughput efficiency. For maximum throughput efficiency, the non-information signals within each transmitted data block must be kept to a minimum and the intervals between successive blocks need to be minimised in both periodicity and average duration. Within limits, an improvement in throughput efficiency can be achieved by increasing *block size*; typical network characteristics limit the increase in efficiency possible at the lower end of the block-size scale, and very large blocks take too long to be retransmitted in the event of error. The optimum solution for any particular network is inevitably a compromise but, in general, data blocks tend to be much longer than in a packet-switched network. All of these factors have a direct influence on overall network cost and need to be taken into account when choosing both the network-interface hardware and the means of interconnection.

The network designer's objective is to achieve an acceptable average response time at minimum overall network cost. The most appropriate transmission speed must be determined for each link independently, on the basis of the forecast data traffic flow generated by the terminal or group of terminals it will serve (Section 6.5.2: see the example for a data-network link). Network-interface equipment must then be selected to match the proposed computer and terminals to the chosen speed. The available range of leased private circuits offers a variety of transmission speeds which will permit the forecast traffic to be carried with varying degrees of efficiency; depending on the speed required, each link will be either an analogue circuit/pair of circuits or a digital circuit (Section 3.1.1). For a given data traffic flow, a slower transmission speed equates to lower leasing charges and more efficiently used circuits, but increases the probability of delays during busy periods and hence the average response time.

Telex network stations are star connected to each switching centre in the manner of a telephone network. Minimum overall network cost is the normal criterion for optimising the number of switching centres and minimum cabling cost their location within each catchment area, as described for a telephone network in Section 8.1. However, because of the smaller network size, practical considerations tend to override purely economic factors in determining the number of switching centres, particularly for private networks. Switching centres are interconnected by a mesh-type network, normally partially connected as for a telephone network; the greater efficiency of a hierarchical network structure makes this increasingly justifiable with growing network size.

Any of the methods described in Section 8.1 for routing traffic in a telephone network can be applied to a message-switched store-and-forward type network, and the same constraints and advantages apply. The implementation of dynamic alternative routing, available only with modern stored-program-control switching systems, is simplified by the fact that messages carry their own routing information introduced at the originating terminal, and hence do not need to be guided through the network by the separate signalling of routing digits.

Messages are individually switched, separately in each direction, and do not necessarily result in an immediate response, so traffic flow is expressed in terms of messages instead of message pairs. Since messages are switched separately and switching-centre message handling time is an additional delaying factor, response time is significantly longer than for an unswitched network. Speed of operation is, however, less important where the requirement is simply to exchange messages between telex-type terminals. Lower transmission speeds offer circuit economies to the network operator by allowing a number of links to be multiplexed over each circuit; this is a useful means of enabling the private-network operator to reduce leased public-network circuit requirements. The buffering effect of store-and-forward working also permits transmission-speed changes to be made across a switching centre. This permits a wider choice of terminals to interwork, in particular allowing obsolescent low-speed machines to work satisfactorily to modern higher-speed terminals without the need for additional user equipment or intervention by the users concerned.

Instead of the best possible response time of an unswitched network there is a unidirectional *best possible message transit time,* which includes the message handling time per switching centre as well as link transmission times. The fact that routings differ both in length and in the number of switching centres involved means that users experience a range of possible message transit times in practice. The store-and-forward capability introduces an additional variable element into switching-centre message handling time, which varies according to the link congestion encountered on outgoing circuits and thus depends on the link capacity relative to forecast traffic flow. An *average message transit time* is, therefore, estimated on a probability basis.

A message-switched network permits greater traffic-carrying efficiency per circuit and hence lower network costs than would be possible with circuit switching. To obtain these savings, circuits must be operated at a poorer grade of service (GOS) and thus an improved occupancy, so that fewer are needed to carry a given amount of traffic. However, reduced costs are achieved at the expense of message transit time, which increases sharply when links become too congested. The network designer's objective is to obtain a balanced and optimally efficient solution which loads the linking circuits to the maximum possible occupancy whilst allowing an acceptable message transit time to be met. Given a typical modern switching-system design based on fixed-capacity subsystems, as described in Section 7.3, link capacity – influenced by both GOS and transmission speed – is the only real variable the designer can manipulate in order to achieve this.

The criteria for deciding the optimum number and location of telephone switching centres in a network apply equally to a *packet-switched data network* but, as with a message-switched network, economic factors tend to be less important than practical considerations. A hierarchical network structure becomes more efficient than a simple mesh arrangement as a network grows in size. Routing options do not arise because dynamic alternative routing is integral to the packet-switching concept; however, before choosing a switching system the network designer must select either *datagram* or *virtual-call* working (Section 3.3).

Data terminals, mainframe computers and printers can all interwork without restriction over a packet-switched network. Throughput efficiency, as discussed for an unswitched network, is of equal concern to the packet-switched network operator, but it should be remembered that error control is on a *link-by-link* basis and involves the retransmission of much shorter data frames. Cross-network delay is comparable with that between terminal and computer across an unswitched network and very much less than that between terminals across a message-switched network. Delay is normally expressed in terms of an *average response time*, as for an unswitched network, but also taking into account the variability of both switching-centre packet handling times and alternative cross-network routings.

The modern subsystem design concept allows only limited scope for the precise dimensioning of switching centres (Section 7.3). Digital links permit much higher transmission speeds in a packet-switched network, those between switching centres operating at up to 64 kbit/s; the speed-changing facility across a switching centre allows a range of lower terminal operating speeds to be accommodated. As in the case of a message-switched network, it is possible to minimise network costs by optimising link loading; this involves choosing the most suitable transmission speed and circuit occupancy (see the packet-switched network example in Section 6.5.2) within the constraints imposed by an acceptable average response time. In practice, however, the problem of accurately forecasting traffic loads on a link-by-link basis when there is dynamic alternative routing, leads public-network operators to adopt the safe policy of designing to a less-than-optimum circuit occupancy. It is possible to achieve greater network efficiency where the overall traffic level is more predictable, as may be the case in a private network, although this may require a process of iteration over a period of network-wide traffic measurement to achieve optimum results.

8.3 Local and wide area networks

The *baseband local area network* (LAN) was specifically designed for the high-speed error-free transmission of packetised data over small private networks. It combines the network simplicity and shared transmission-path of an unswitched data network with the flexibility of a fully fledged packet-switched network; however, the area and number of stations that can be served are restricted by transmission distance and capacity limits. The shared transmission path, configured as either a ring or a bus, is provided by a single easy-to-install cable, along the length of which the stations – terminal, computer or printer – are connected at intervals, either directly or via short spur cables (Section 5.3).

There are many different proprietary versions of baseband LAN available within the broad classifications referred to in Section 5.3, each supplied as a complete system. Each system provides an exclusive range of equipment and facilities and often uses unique versions of the basic protocols to control the packetisation of data, switching, error correction, etc. Each system is also uniquely limited in respect of the maximum length and type of its main and spur cables, and the extent to which repeatered branching connections are feasible to extend the area of coverage; in some cases stations and spur cables can only be connected to the main cable at precisely specified intervals. The main-cable shared transmission path is limited in capacity by the speed at which it operates, as is a link in an unswitched network, and the fact that each system operates at a particular fixed transmission speed imposes on it an overall capacity limit in terms of the number of stations that can be served and their individual transmitting speeds. These various criteria also determine the efficiency of net data throughput. The performance of each proprietary system is, therefore, constrained by a unique range of limits which determine its suitability for a particular installation.

The network designer's task is to select a baseband LAN system which is capable of meeting, at minimum cost, the requirements of the particular installation under consideration, both initially and in the future. With ring-based systems, a basic choice has to be made between those which employ empty-slot and token-passing data-transfer techniques (Section 5.3); in general, the so-called *slotted rings* offer a greater capacity and transmission-speed capability, whereas *token rings* are simpler and cheaper to implement. The choice of system can be narrowed by considering those performance limits where the limiting effect is readily apparent; for example, the size of area that can be served. System capacity, however, can only be considered in relation to *cross-network delay* because, as with any other data network, this increases the more heavily a LAN system is loaded. The most critical data transactions in this respect are those which involve interactive working between a terminal and a computer; hence this delay is expressed in terms of an *average response time* as for an unswitched network. This overall average response time is measured, as always, from terminal input to receipt of an answer; this includes a component representing the average response time of the computer(s) associated with the LAN. It is only the remaining components of overall average response time, covering the transmission of data packets to and from the computer, which are influenced by the LAN system itself. Baseband LAN systems incorporate buffering between a station and the transmission path, to allow messages to be input without the assurance of immediate transmission and to permit changes of transmission speed. In addition, some systems cater for repeated attempts to transmit and, where there is a ring-based transmission path, the continued circulation of data packets waiting for the destination station to become free. A number of different variable elements of delay affect the average response time at peak periods, therefore, and these can only be calculated on a probability basis. The complexity of these calculations has led manufacturers to produce computer packages which determine the capacity of a LAN for a particular average response time, although the availability of these packages is sometimes severely restricted.

The individual channels of a *broadband LAN* (also described in Section 5.3) must be considered separately, the approach to network design depending on the service for which each is to be employed. For particular mixed communications needs, the flexibility of being able to use the channels in different ways offers enormous scope to the network designer; however, the many different factors to be considered make the design process both complex and lengthy if a cost-effective solution is to be achieved. A channel to be used for telephony must be considered as a separate telephone network; a channel to be used as a point-to-point data link must be treated as an unswitched data network; and a channel which will carry data traffic using a baseband-LAN-type access technique must be designed on the same basis as an equivalent baseband LAN.

Wide-area-network planning concentrates on the means of interconnecting the various LANs, each of which is, effectively, a self-contained entity. Inter-LAN response time is an important new factor to be considered. Before detailed design work on the LANs and inter-LAN links can commence, however, a fundamental decision has to be made concerning the method of interworking to be used, since this determines the type of terminal equipment required for the LANs and the link capacity needed. The initial choice is between *point-to-point* or *broadcast* working (see Section 5.3), with the latter offering simplicity at the expense of increased link capacity and hence being more suitable for smaller and less complex networks. If point-to-point working is selected, it is then necessary to make a further choice between *router* or *data-link bridge* connection (also explained in Section 5.3). Routers and bridges each have their particular technical merits, and data communications engineers tend to be partisan in their views on which can be expected to offer the best solution for a particular network configuration. The most important consideration from the network operator's point of view, however, is almost certain to be overall network cost. The effect of the choice(s) on cost is not immediately apparent because each alternative offers more simplicity and less capacity in a different area at the expense of greater complexity and more capacity elsewhere. The choice should ideally be resolved, therefore, by a broad-brush comparison of the costs involved when each alternative is applied to the particular network requirements under consideration; this should include running as well as initial costs, since the former may be influenced by differences in leasing charges for the inter-LAN links.

Chapter 9

Traffic-carrying performance evaluation

9.1 The role of the performance engineer

Throughout this book, emphasis has been placed on the direct relationship between the traffic-carrying efficiency and the cost of a telecommunications system or network. The *traffic-handling performance* of every system and network is limited by its configuration, the concepts employed to transmit, switch and route traffic, and the design of the manufactured products which implement those concepts. To achieve a given standard of performance, therefore, requires careful attention to system choice and network design. The situation can arise where the required performance standard is beyond the technical capability of a proposed system or network solution, giving rise to the need either to find an alternative solution or to accept a lower standard of performance. Traffic-handling performance can be estimated by a network designer with sufficient accuracy for most dimensioning/sizing purposes, but there will be an occasional need for a detailed evaluation of system/network performance by a specialist performance engineer. Telecommunications managers should, therefore, have an understanding of what can be achieved in this field.

The mathematical analysis of telecommunications traffic behaviour pioneered by Erlang and others (Section 6.2) has become a recognised subject of academic study throughout the world under the title of *teletraffic theory*. The objective of its practical application, known as *traffic engineering*, is to achieve optimum traffic-handling efficiency at minimum cost. To this end, system/network performance is analysed under various postulated traffic conditions; this enables designs to be improved and optimum dimensioning rules to be established. Similar mathematical techniques to those used in teletraffic theory are applied to the investigation of other areas of performance throughout industry, but their application to problems encountered in the design and sizing of mainframe computers and their associated networks, and to the prediction of electronic-system reliability, is of particular relevance. This work is referred to in the computing field and elsewhere as *performance engineering*. The convergence of computing and telecommunications technologies during the last two decades has meant that the studies of system and network behaviour in both areas now have much in common. Although the two areas of study are, in the main, still pursued quite separately, the benefits to be gained from a cross-fertilisation of ideas and the use of common terminology are increasingly recognised. In particular, there is a growing acceptance among telecommunications engineers that 'performance engineering' is now the more appropriate term to describe the practical application of teletraffic and related performance theory to modern computer-controlled telecommunication systems and networks. It is in this sense, as applied to both telecommunications and computing practice, that the term 'performance engineer' is used here.

The increasing demands on teletraffic theory since the early days of manually switched telephony have an important bearing on the present-day role of a performance engineer. Originally,the only concern of teletraffic theorists was to measure and forecast traffic flow, in order to develop a consistent method of calculating how many circuits should be provided throughout a network. The automatic switching systems that succeeded manual switchboards had a highly modular switch structure, with distributed control implemented through the interaction of relatively slow-acting (by modern standards) electromagnetic circuits. The need to evaluate the efficiency of various switch configurations placed new demands on teletraffic theory, and dimensioning methods had to be evolved for the wide range of components required to construct a switching centre. The eventual move towards a more centralised form of control introduced the concept of queuing – for example, of call attempts during set-up. This required new mathematical processes to be developed to enable the prediction of system behaviour. The most recent systems incorporate high-speed electronic circuits and control by computer software. These changes and the resultant equipment miniaturisation have led to the need for fewer equipment modules, each having a wider variety of tasks to perform; this has reduced dimensioning flexibility and simplified the dimensioning process. System operation, however, has become more complex; many different software processes and circuits have to interact with microsecond timing, making performance evaluation more difficult. It is no longer possible to analyse the performance of a switching system by mathematical techniques alone, and system behaviour has to be simulated on a computer as described in Section 9.2.

These changing telecommunications switching-system concepts have widened the scope for performance engineering involvement. In the early days, the chief concern of a traffic engineer was how to configure and dimension the components of a system or network to achieve the best overall traffic-handling performance; there was little need to analyse the operation of individual components. Now, the overall performance of a system, and thus of a network in which it is used, is limited by the traffic-handling capability designed into the system components. This is a direct parallel with the situation that obtains for computer systems and networks. In order to optimise the traffic-handling efficiency of a system, therefore, the present-day performance engineer must be able to influence the way in which the electronic circuits and software behave and interact. The well known 'feedback' approach to solving problems involves following a repetitive cycle of (a) testing a possible solution and (b) using the information obtained to improve the solution-finding mechanism, until the best possible result is obtained. This feedback technique is already widely used throughout manufacturing industry where, for example, the results of performance evaluation measurements are used to correct design weaknesses in new prototypes before they reach the production stage. The practice can, however, be applied even more cost-effectively when it is possible to analyse the initial design theoretically and predict the results that will be obtained. The performance engineer is able to do this, predicting the performance of a new design 'on paper' and thus allowing its capability to be optimised before manufacture commences. Prediction is, in fact, the basis of most telecommunications and computing performance engineering, since it is seldom possible to control the traffic conditions on live equipment sufficiently to provide meaningful measurements of system/network behaviour. If the traffic-handling performance of a telecommunications or computing system is to be assured, therefore, a detailed performance engineering involvement is essential at a *very early stage* in the design of the electronic and software system elements. This need is particularly relevant to integrated-circuit design, where progress has been such that quite complex circuits can now be contained on a single 'chip'. Ideally, the traffic-handling aspects of any such circuits need to be considered at the conceptual stage of chip design, since the costs of redesign may well preclude corrective action at a later stage.

Telecommunications and computer equipment suppliers throughout the world have shown a reluctance to engineer their products for optimum traffic-handling performance to the extent that is desirable. Performance engineers tend to be employed in a consultancy role, even by some of the leading manufacturers of mainframe computing systems and telecommunications switching systems; this often leads multinational companies to centre their expertise in one country. These engineers advise on the dimensioning/sizing of finished products and the traditional network aspects of configuration and traffic routing, but often do not have a sufficient degree of detailed involvement in circuit/software design. Some smaller companies do not have access to specialist performance engineering advice. It is advisable, therefore, to seek evidence of performance evaluation in respect of published performance figures, especially when the product in question has no history of proven effectiveness in operational use.

Performance engineers are also employed by the principal public-network operating organisations throughout the world, again largely in a consultative/ advisory role, and by major consultancy firms. These engineers only become involved in the detailed system design process when, exceptionally, a product is designed wholly within a public-network operating organisation or in partnership with one or more manufacturers. Their main task is to establish the capability of production systems being purchased for use in or attachment to a network, and to advise on the use of new network concepts and network development generally. Performance engineers also undertake specific network design tasks and develop computer aids for use by non-specialist staff for evaluation, design and dimensioning purposes.

The role of the performance engineer can, therefore, be most easily summarised in terms of the different performance engineering objectives in the two main areas of system and network design. In *system design*, the objective is to guide and influence the software and circuit designers at the earliest possible stage, in order to ensure a finished product that will transfer and/or switch telecommunications or computing traffic as efficiently as possible, at minimum cost. If the performance engineering aspects of system design are ignored, or considered at too late a stage, this inevitably leads to the use of inefficient circuitry and/or software in production systems and is a contributory cause of system failures under adverse traffic conditions. In *network design*, the objective is to ensure the transfer of information between users by the most efficient means possible, in terms of the configuration and dimensioning/sizing of network components and the routing of traffic, at minimum cost. If the performance engineering aspects of network design are ignored, the result may well be a network that fails to cope with the demands placed upon it and/or costs more than necessary.

The increasing use of *computer-aided design* (CAD) techniques, intended to speed system design by ensuring a uniform, structured approach to electronic and software design problems, offers the opportunity of incorporating performance-optimising principles into the design of more new products. The CAD concept envisages that small teams of experienced design engineers will undertake the development of computer-based design aids for the benefit of those engaged in product design. By ensuring that these design aids take performance optimisation into account, and eventually integrating performance prediction/ feedback processes within the overall CAD process, it should be possible to ensure that all products designed in this way perform with maximum efficiency. The performance engineer of the future, therefore, can be expected to specialise in the development of the performance evaluation elements of CAD processes and other computer aids intended to simplify the performance evaluation of networks.

9.2 Performance evaluation methods

When information is transferred by a telecommunications or computing facility, the transfer is achieved as a result of several chains of automatically controlled actions, each chain being triggered off by the initiating user or by an action of a preceding chain. These actions are referred to in computer parlance as *events*. Apart from those events which are caused directly by a user's actions, the control of events at any particular stage of the transfer process is always centred on a *system*. In order to evaluate the performance of the system or network concerned, therefore, the performance engineer must first know how these events are caused to happen, what their effect is, and precisely when and in what sequence they occur after they have been triggered off; this requires a detailed study of the software processes and circuit functions of the controlling system(s). A system analysis to this depth can be very time consuming and, since all the details required for *performance evaluation* have to be established during the design process, can be carried out more effectively from within a design team than outside it. When the evaluation is being undertaken independently, much depends on the ability and willingness of the manufacturer to supply the details required.

Performance evaluation in telecommunications and computing is mainly predictive, although measurement is used to a limited extent. The *measurement* facilities integral to a system provide the information needed for changes to be made to the dimensioning/sizing of the system itself and the network in which it is being used, in order that changing traffic demands can be matched by appropriate equipment quantities. These standard facilities, however, are rarely adequate for performance evaluation purposes; for example, although measured response times are an important indicator of the day-to-day performance of unswitched and packet-switched data networks, they do not meet the objective of performance evaluation, which is to determine the *limits* of performance. It is seldom feasible to take additional measurements on a live operational system or network; not only do non-standard measurements necessitate the time-consuming development of special measuring equipment but, in addition, great care has to be taken to ensure that system operation is not affected by the measuring process. Even if an adequate range of measurements is practicable, the instantaneous variation of traffic load on most operational systems and networks precludes measurements at the fixed traffic-load values necessary to establish performance limits.

For these reasons, the performance of an *operational* system or network can seldom be adequately evaluated by measurements alone and, in the absence of a predictive evaluation, optimisation has to be confined to dimensioning/sizing aspects within the established configuration. Nevertheless, when it proves impossible to obtain enough detailed information on system operation for a predictive evaluation, there is no alternative to the measurement approach if performance claims are to be tested. The only satisfactory solution in this case is to measure the performance of a live but *non-operational* system which has been loaded with a predetermined level of artificial traffic. *Artificial-traffic generators*, which are designed primarily for the commissioning of newly installed systems and make use of a computer to simulate large numbers of users, are employed for this purpose. System measurements may also be taken, for example, to check that the performance of an artificially loaded prototype system is in accordance with earlier predictions, or to obtain detailed information for input to a predictive evaluation of future system development proposals.

The techniques used for *predictive* evaluation can be divided into two categories, analytical and simulation. The usual approach to a predictive evaluation is to break down the overall system or network concept into several self-contained areas, so that the operation of each can be evaluated independently in order to determine the behaviour of the whole. For a switching system these areas might include the call-control process, processor overload control, switching element, line-termination element, signalling, routing, etc. The most appropriate technique, analytical or simulation, can then be selected for application to each area.

The *analytical* approach requires a selective study of the event sequence to determine whether it is possible to represent system or network behaviour adequately in mathematical terms. The aim is to construct a *mathematical model* which is a representation of the *essential* traffic-handling effects of system/network behaviour; in effect, the model is a mathematical précis of the real sequence of events. This requires equations to be selected and used in such a way that their effect on a hypothetical traffic-load input figure will closely resemble the effect of the real system or network on real traffic. The solution of these equations then provides a measure of the efficiency of traffic-handling performance. In this way, performance can be evaluated for a range of traffic loads and the limits of performance established. An experienced performance engineer can swiftly determine whether an analytical evaluation is feasible and, where the technique is readily applicable, can produce useful results relatively *quickly* and *cheaply*. However, it can prove difficult or even impossible to use this technique to model very complex system behaviour with sufficient accuracy. The technique is normally applied, therefore, to aspects of system or network operation (a) which are sufficiently straightforward to permit modelling to an acceptable degree of accuracy, or (b) where the expected savings in time and cost are judged to outweigh considerations of accuracy, as during a conceptual design phase.

The second of these predictive evaluation techniques, *simulation*, involves the writing of one or more computer programs to represent the event sequence, but differs from the analytical approach in that the representation is not limited by the availability of suitable mathematical processes and can be as precise as the performance engineer requires. The aim is to construct a *computer model* which is, in effect, a computer-language précis of the real sequence of events. Specially designed or adapted languages are used for this purpose and the programming is carried out by the performance engineer. Traffic load is made a variable input to the program(s), which are run on a computer to establish what happens when the load and various performance-influencing characteristics are changed, as with analytical evaluation. Great care has to be taken to ensure that the real events are accurately represented, and the programming task is inevitably *time consuming* and *costly*. The programs for the evaluation of a public-network switching system are likely to be particularly complex and lengthy, requiring the power of a large mainframe computer to run them. Once completed, however, a simulation program can provide almost limitless scope for establishing the extremes of performance with great accuracy. The great advantage of the simulation technique is, in fact, its capability of accurately representing the behaviour of the most complex system or network, and it is ideally suited to the predictive evaluation of detailed system designs. Because of the time needed to produce an effective simulation, however, work undertaken in parallel with system design needs to be closely integrated in the design process and tightly controlled to ensure that design deadlines are achieved.

The simulation technique, being computer based, is ideally suited to auto-misation within the CAD process referred to in Section 9.1. At present, simulation programs have to be written in a different computer language from system-control programs, each type of language having been developed specifically for its particular purpose. Events that are relevant to performance have to be selected from the control program and rewritten in a simulation language before the evaluation of the former can commence. What is needed is a means of identifying, extracting, assembling and evaluating these performance-affecting events automatically during the course of control-program design. If all system design programs are structured in a standard manner, the automatic translation of selected events from a control-program language to a simulation language should become feasible as a first step towards this goal. Ultimately, the ideal solution would appear to be to develop a common language that is suited to both purposes. It might then be possible to develop a system design procedure which guarantees an acceptable standard of performance, by means of an automatic iterative process of (a) predicting performance directly from the control program under design and (b) feeding back information on any short-comings to initiate design corrections.

Chapter 10

The effects of traffic-carrying performance and public-network tariffs on private-network costs

10.1 Traffic-carrying performance and public-network tariffs in the context of private-network costs

The term *traffic-carrying performance*, as applied to any telecommunications network, is used to describe how effectively that network is able to transfer information between users. Earlier chapters have shown that traffic-carrying performance is inversely related to the percentage of calls lost and call set-up delay experienced with a circuit-switched telephone network, and to information-transfer delay across all types of data network. Call losses or delays can only be reduced by providing more or, with data, faster circuits and more equipment, at greater expense; thus traffic-carrying performance is also directly related to overall network cost. By matching the traffic-carrying performance of a private telecommunications network as closely as possible to the needs of the business concerned, it is possible to ensure that network costs are no greater than absolutely necessary. This objective can be achieved by designing the network to meet a *performance target*, as described in Sections 10.2 and 10.3.

It must be remembered, however, that the calculated traffic-carrying performance of a network relates solely to the sufficiency of system equipment and circuits provided to carry the expected level of traffic, on the basis of what amounts to an assumption that all the equipment and human beings involved will always function correctly as and when required. A performance target cannot take account of communication failures arising from (a) traffic levels in excess of those forecast, (b) user's equipment faults, human error or inability to respond, etc., or (c) network equipment faults. These factors will all have an adverse effect on the actual performance experienced when using a network. The network operator, although able to take action in respect of (a) to match increased traffic levels once they have been detected, has little influence over the users or their equipment responsible for (b). The increased reliability to be expected from modern telecommunications systems lessens the probability of network equipment failure (c), but faults can and do occur, affecting both switched public services and leased private circuits. Power failures, although rare, are another possible cause of a breakdown in communication. The possible effects of such breakdowns on a business should at least be considered. A completely safe network is not a realistic possibility, but safeguards in the form of alternative routing schemes and emergency power supplies (standard for public networks) can be built into any private telecommunications network at additional cost if the importance of communications to the business is considered to warrant this.

Most private-network operators have recourse to public-network services, sending some of their traffic over one or more public switched networks and/or leasing public-network circuits. The charges levied by public-network operators for these services have an effect upon private-network costs that is quite separate from traffic-carrying performance considerations; the bases on which these public-network tariffs are determined and applied are, therefore, examined in Section 10.4.

Above all, the private-network operator will wish to be assured that the most effective and least costly system/network solution is being provided for each task, taking into account traffic-carrying performance, public-network tariffs and, where appropriate, network security. Finally, therefore, the process of selecting the right option from a number of technically suitable alternatives is discussed and illustrated by worked examples in Chapter 11.

10.2   The design of private telephone networks to a performance target

The traffic-carrying performance of any telephone network and its switching centres is determined by several different and sometimes conflicting performance criteria, applied at a number of separate points; all of these must be considered and balanced to achieve the desired overall result. More capacity in the right places increases the probability of establishing a successful connection, but this reduces traffic-carrying efficiency and increases cost. The cost-conscious private-network manager is, however, more likely to be interested in the corollary of this statement, that less capacity increases traffic-carrying efficiency and reduces cost, but reduces the probability of establishing a successful connection. Reducing the probability of successful call set-up also inevitably increases the average time taken to set up a call; this, of course, always includes a fixed, minimum element of time which depends on (a) the number of switching centres involved in the call and (b) the switching and signalling capability of the chosen switching system(s). Ideally, the aim should be to limit costs to the point where call set-up performance is just acceptable.

The manager of a private telephone network can influence the performance and thus the cost of that network by agreeing, with the network designer, a *performance target* which will indicate the *balance required between capacity and call set-up*. If the network designer is given no guidance in this respect, the network is likely to be dimensioned to standards derived from generalised past experience, not necessarily appropriate to the particular needs of the business concerned. Provided the designer has been given adequate and accurate traffic and connection forecasts on which to base the design, it is unlikely that the ensuing performance will be found wanting. It is very important, however, to ensure that the standards applied do not prove to be excessively stringent for the business needs of the network operator, incurring unnecessarily high network installation and running costs (including circuit leasing charges). Furthermore, unless the standards employed by the designer are disclosed, the manager will not be in a position to judge whether the network is performing as intended.

Overall grade of service (GOS) (Section 6.5.1) can provide a broad-brush but nevertheless convenient means of setting a performance target for a private telephone network. Upgrading the overall GOS increases the probability of successful call set-up but reduces traffic-carrying efficiency, thereby increasing costs; degrading the overall GOS reduces costs. As with traffic-carrying performance criteria generally, overall GOS is concerned only with the sufficiency of switching-system equipment and circuits relative to the forecast traffic level.

The overall GOS across a telephone network takes into account the GOS across each switching centre and the GOS applied to the dimensioning of each interconnecting traffic route. The GOS offered by a modern electronic switching centre or PBX, whether analogue or digital, is to a large extent determined by the design of the switching system. Provided the switching centre or PBX is dimensioned (a) in accordance with the published instructions and (b) to a traffic forecast which proves to accord closely with the level and balance of traffic flow eventually experienced, the manufacturer should be able to quote a figure for the average end-to-end GOS across it. The question of *traffic balance* is particularly important, since the performance of a switching system can deteriorate rapidly if, for example, one particular number or traffic route is in persistent demand and this has not been taken into account in the dimensioning process. The GOS offered by a correctly dimensioned switching centre or PBX subjected to the forecast traffic demand is, therefore, largely fixed by the system choice and cannot subsequently be varied to any significant event. (For this reason it would clearly be inappropriate to apply an overall GOS target figure to, for example, a single PBX.) This switching-system GOS figure can be expected to be in the range 0.01 to 0.001, the latter being typical of a well designed public-network-type system. The GOS applied to a traffic route, however, can be varied over a wide range, and it is in this area of network design that the traffic-carrying performance of a network is optimised.

When deciding on an overall GOS target figure, it can be helpful for the non-technical manager to consider the outcome of choosing a particular target positively, in terms of what is effectively the reciprocal of GOS – that is, the percentage of *successful call set-ups* that can be expected during an average busy hour; thus a GOS of 0.1 equates to a 90% success rate. It is important to remember that the busy hour to which it will apply represents a worst-case situation; performance will always be better outside that busy hour unless some exceptional and unforeseen event occurs which degrades the network through equipment failure or causes an unusually high level of traffic flow. The aim should be to ensure that the overall GOS target is no more stringent than absolutely necessary. The precise relationship between overall GOS and network cost varies from network to network and is not immediately apparent, but it is possible to judge the increased cost of an improved standard of service by seeking approximate quotes for *two different overall GOS target figures*. It should be borne in mind, howerver, that this practice may well incur extra design charges.

To be of use to the network designer, an overall GOS target must be related to a specific date in the future. In the normal situation of an expected growth of traffic, the actual overall GOS experienced will progressively deteriorate over time until network capacity is augmented. It is, therefore, usual to align the overall GOS target to the date to which the traffic forecast applies, that is, the planned design date of the initial (or extension) equipment installation when further network augmentation is expected to become necessary (see Section 6.5.1 and Fig. 6.2).

Section 6.5.1 explained that overall GOS cannot be precisely quantified because it is the collective result of individual GOS standards, each based on an element of probability and applied separately to the component parts of a network. The GOS experienced at each stage of a call attempt will only be worse than the GOS standard applied to the network component at that stage if the actual traffic flow exceeds the designed level. In any other circumstances the actual GOS will be better than this standard, but it can require considerable mathematical expertise to estimate how much better. The process of estimating the overall GOS of a network at a particular time is made more complex when (a) the network components have been augmented at various times, causing each to be at a different stage between its periodic augmentation dates, (b) the timing of the busy hours to which the GOS standards relate varies between network components, (c) the actual traffic levels are at variance with those forecast, or (d) alternative routing offers a choice of possible paths through the network. All of these factors apply to a large public network. An additional complication arises because these networks are normally configured as a hierarchical structure of tiered switching centres (Section 8.1), with cross-network traffic concentrated on to a small number of large (trunk/toll) routes. Cost savings are effected by dimensioning these more efficient large routes to a relatively poor GOS compared with that applied to the smaller routes on the periphery of the network. Thus a range of different GOS values is applied to different traffic routes, and the number of routes involved on any particular call will also vary, depending as it does on the relative locations of the calling and called users. A network-wide overall GOS figure, therefore, can only be estimated in terms of either an average value or a worst-case value, and the complex task of estimation is normally undertaken by specialist performance engineers.

These difficulties in estimating a realistic overall GOS figure apply equally to large private networks, and the reverse process of correctly apportioning an overall GOS target between the components of such a network requires specialist performance engineering knowledge or advice beyond the scope of this book. However, most private networks are relatively small in comparison with a public network and are usually dimensioned to one *common busy hour*. Where this is so and *single-choice routing* is employed throughout the network, it becomes possible to estimate a meaningful worst-cast overall GOS by simply adding the individual GOS standards applied to successive network components encountered along the longest possible path through the network (Fig. 10.1a).

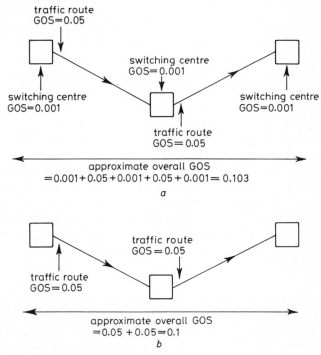

Fig. 10.1 *Overall grade of service:*
(a) Concept
(b) Excluding switching-centre GOS where this is very much better than traffic route GOS: switching-centre GOS = 0.001 as in (a).

It is important to remember, however, that such an estimate is only valid at one time, just prior to network augmentation, if (a) all the component parts of the network are being augmented concurrently and (b) the actual traffic flow at that time is as forecast. Unless the traffic flow proves in reality to be in excess of that for which the network has been designed, the actual overall GOS encountered can never be worse than the estimated figure; in most circumstances it will be better. The estimated overall GOS, should therefore, be regarded as no more than a guide to the probable worst-case situation. If the overall GOS target figure is similarly treated as a guide to the worst-case requirement, the individual GOS standards applied to each network component can be apportioned so that they add up to this overall figure.

In practice, as mentioned earlier, the scope for significant adjustments to GOS standards is limited to those applied to the dimensioning of the interconnecting traffic routes. Where, as is usually the case, the GOS offered by a switching centre or PBX is very much better than that needed to dimension the traffic routes, the former has little impact on the overall figure. It is then acceptable to exclude all switching centres and/or PBXs from the calculation and apportion the overall GOS target figure between the GOS standards applied to the successive traffic routes across the network (Fig. 10.1b). The apportionment process is simplified if a common standard is applied to every traffic route; in the example shown, with a maximum of two routes in tandem, each route would need to be dimensioned to a GOS of 0.05 to achieve a typical overall GOS target of 0.1. Unless the amount of traffic has been seriously underestimated, this equates to *at least* a 90% chance of successful call set-up in the busy hour and a very much better chance than this at any other time.

To achieve the same overall GOS in a smaller network which has two switching centres and a single interconnecting traffic route, this route would only need to be dimensioned to the less stringent GOS of 0.1, equal to the overall GOS target figure. On the other hand, a network across which a maximum of four switching centres and three interconnecting traffic routes may be involved in a call would, on the same basis, require each route to be dimensioned to a better GOS of 0.033. Where, as is normally the case, such an intermediately switched network is partially connected, some traffic will be routed indirectly over routes which also carry directly routed traffic. Where this is so, the presence of indirectly routed traffic on a route takes precedence in determining the GOS to be applied; this means that directly routed call attempts, which involve only one route, will have a greater probability of successful completion than indirectly routed call attempts over that same route. It can be stated, therefore, that the more intermediate switching centres a call attempt is able to encounter successively across a network, the better is the GOS that must be applied to each interconnecting traffic route to give the same overall standard of service. Although an improved route GOS implies the need for more circuits, more switching centres mean a large network with larger, more efficient routes, of which each circuit is able to carry an increased amount of traffic.

The overall GOS target concept can be readily applied to any wholly private-circuit routes and to those high-usage private-circuit routes dimensioned as described in Section 8.1. Where traffic to a particular destination is routed wholly or partly via a public network, however, the GOS target is only applicable to that routing in so far as it influences the number of local circuits required to access the public network from each PBX or switching centre (see the example included in Chapter 11).

10.3 The design of private data networks to a performance target

An *unswitched* data network, having no network facility for switching or storing messages, relies on the time sharing of data links between its terminals and a central main frame computer, and performance is measured in terms of an average response time (Section 8.2). The network designer's task is to choose and size the computer, select the most appropriate network configuration and access arrangements, allocate the stations/terminals to links and optimise link transmission speeds (Section 6.5.2), to achieve an acceptable performance at minimum cost. The optimisation of throughout efficiency and hence average response time is a task to be undertaken *after* the major decisions on the network and its equipment have been finalised, but the network manager can influence these earlier decisions by agreeing, with the designer, a *target average response time*. This target should apply to a defined date in the future when computer/network usage is expected to match the capacity of the equipment provided. A faster response time equates to increased installation and running costs (including circuit leasing charges); hence it is important that the target figure should be no more stringent than absolutely necessary. The economic consequences of a faster response can be revealed, at the possible expense of increased design charges, by considering *two different target response times*.

The failure of newly commissioned computer networks as a result of over-loading is by no means uncommon. Although this sometimes occurs because processing requirements have been underestimated, by far the most common cause of overloading arises from the enthusiasm of users who wish to experiment with the new system in order to gain expertise, thus generating a much higher rate of terminal usage than would be normal. To overcome this problem, both the computer and its network would need to be over provided for normal usage and would be inevitably more costly, although where a growing network demand is anticipated it could be argued that the additional expenditure would merely be brought forward in time. Only the network operator can decide whether increased cost is justifiable to avoid the possibly damaging media publicity and undermining of users' confidence in the new system which could result from system failure. A precise specification of the network operator's requirements in this respect is, therefore, essential.

The traffic-carrying performance of a private *message-switched* data network is influenced by the same factors that apply to a telephone network, but the inclusion of a store-and-forward facility adds a new dimension to these. The ability to store messages in transit enables the network to accept messages without the need first to establish a communication path between the calling and called users, and introduces an additional network delay factor which has to be taken into account at the network design stage. It also enables circuits to be more heavily loaded and thus used with greater efficiency. Although the concept of an overall GOS target figure could be adapted to a private message-switched data network, in practice it is usual to consider message delay as the sole criterion of network performance, in terms of a *target average message transit time*. This target should be applied, as in the case of a telephone network, to a defined date in the future when the traffic flow is forecast to be at its maximum level prior to network augmentation. Network configuration generally follows telephone network practice and, since link capacity between switching centres is the only readily adjustable variable that can influence message transit time, it is the dimensioning of these links which is of primary concern to the network designer. When there are more links in tandem, more stringent performance standards need to be applied to the dimensioning of each link in order to maintain a given overall network performance standard, as is the case for a telephone network.

Outgoing link capacity and the message-handling performance capability of the chosen switching system together determine the message transit time across each switching centre (Sections 3.2 and 8.2). Link propagation times are sufficiently short to be ignored; hence it is possible to calculate the permissible message transit time per switching centre with sufficient accuracy by simply dividing the overall target figure by the maximum number of switching centres that may be encountered in tandem across the network. Link requirements in terms of the number of channels, multiplex facilities and leased-private-circuit transmission speed needed to carry the forecast maximum message flow rate are then determined for each link separately – usually directly on the basis of circuit occupancy rather than in terms of traffic-flow units and GOS – such that the average message transit time through each switching centre will be within the permissible limit. The message-handling performance of modern switching systems is a known quantity, and information on the required occupancy for a given performance standard should be readily available from the manufacturer.

As for a private telephone network, the objective of minimum overall cost requires alternatives to leased circuits – with or without tandem switching – to be considered. It may be cheaper to route the traffic for particular destinations (a) entirely via leased circuits, (b) entirely via the public telex network, or (c) where the chosen switching system makes this a feasible option, via high-usage leased circuits with overflow to the public telex network. The optimum solution can only be ascertained from a detailed cost comparison, using the principles described for a telephone network. Similarly, it is possible to judge the extra cost of better network performance by comparing the results for *two different target mesage transit times*.

Most private *packet-switched* data networks are based on the *local-area-network* (LAN) concept, and their traffic-carrying and computing performance is almost totally dependent upon the correct choice of LAN system and associated main frame computer in order to meet the particular communication and computing needs envisaged. The optimisation of throughput efficiency and hence average response time is then inherent to the design process. Additional considerations only arise where two or more LANs have to be linked in a wide area network configuration or where, exceptionally, a large private network employs public-network-type switching centres. The capacity of the inter-LAN or inter-switching-centre links is then planned on an individual link basis and, as for an unswitched data network, it is desirable to agree a *target average response time* with the network designer. This target performance figure should be applied to a defined date in the future when data transfers and computer usage are forecast to reach the maximum designed capacity level of the installed equipment. Once again, response-time performance is inversely related to cost, and the added cost of improved performance will become apparent if the results of applying *two different target response times* are compared.

There is a wide selection of proprietary LAN systems available and their performance capability is broadly dependent on the mode of operation employed. The first task of the network designer, therefore, is to select the most appropriate mode of operation for the network under consideration. *Baseband* LANs (currently designed for data transfer only, although integrated speech and data communication is in prospect) offer a variety of transmission-path configurations and data-transfer techniques. All *broadband* LANs are configured similarly and have a particular application where, for example, large volumes of high-speed data need to be transferred, possibly using more than one different method of data transfer and with a separate speech-communication capability. Within each category, a number of different proprietary versions is likely to be on offer, each with its particular limitations on data-transfer capacity, distance and the number of users served (Sections 5.3 and 8.3). Ideally, the choice should be made from as wide a range of systems as possible to ensure a least-cost solution. Each system should be evaluated to establish whether it can meet the particular and possibly changing computing and communication needs of the business throughout its operational life, to establish a short list for a comparison of costs. Where a *wide area network* of several LANs is planned, consideration must also be given to the technique and equipment to be used for interconnection and the optimum transmission speed for each link (also referred to in Sections 5.3 and 8.3).

The remarks earlier in this section concerning the over enthusiastic early use of new unswitched data networks are equally valid here. However, since each LAN has a fixed maximum capacity, the problem is only likely to arise when it is loaded close to its limits; any overloading will result in unacceptably long response times. Similar problems of inadequate response-time performance may arise if computing or network usage has been under estimated.

10.4 Public-network tariffs

The operator of a private telecommunications network usually needs to route some traffic directly over a public telephone, telex or data network and, in many cases, lease public-network circuits as an integral part of that network. The tariffs which apply in either case are, therefore, of direct interest if private-network costs are to be minimised. From the public-network operator's point of view, the primary purpose of any telecommunications tariff structure is to price services so as to at least cover the overall cost of providing those services. To enable new services to be introduced and existing services to be maintained and improved, through the funding of research and development and the provision of new and replacement equipment, additional capital usually has to be raised elsewhere. Whether this outside funding is obtained from government or private sources, interest and/or dividend payments are an additional burden which must be reflected in the tariff structure. If the charges are set at too high a level, however, network use will be discouraged to the detriment of the financial well being of the operating administration or company concerned, especially where there is competition. Tariffs also have a very important secondary purpose: they are used to regulate the demand placed by the various telecommunications services on the available network capacity, with the aim of encouraging the use of network resources to the maximum possible extent throughout all the twenty-four hours of every day of the year. This secondary role of tariffs helps towards improving the efficiency with which the fixed assets of an administration or company are used, and is of particular importance in keeping tariffs overall to as low a level as possible.

Tariff structures for public telephone service provide an example of the charging options that are possible. These tariff structures take different forms throughout the world, but essentially all are based on two elements; a fixed periodic standing or rental charge, and a variable per-call charge. The original concept of a *standing charge* was that it should cover the average rental cost of all plant which is exclusive to a user and not, therefore, call related. As a minimum, this takes into account a proportion of the cost of cables etc. between the user and the local switching centre, together with the cost of that element of the switching centre which is provided solely to terminate the user's line. These costs are averaged on a per-user basis for similar types of user installation within the broad categories of business (high calling rate) and residential (low calling rate) user. Where a telephone or other apparatus is rented from the network operator, the rental cost of this is included in the standing charge. Some network operators allow all or a prescribed number of local calls to be made free of charge; the cost of these free calls may then be recovered by an addition to the standing charge, calculated as a fixed amount on an average-user basis, or by a proportionate increase of other call charges. Special categories of call, such as those made for emergency purposes to the ambulance, fire, police and rescue services and to the network operator's repair service, are free of charge to the caller. In all other cases, calls are normally chargeable to the caller on a per-call basis.

Call charges can be based on (a) a fixed charge per call, (b) a variable duration-related charge per call, (c) a variable distance-related charge per call, or (d) a combination of two or more of these possibilities. Calls may be classified in various ways and charged for on different bases, and differing scales of charges may be applied at particular times of the day and on particular days. The revenue from these call charges must cover the cost of those elements of the telephone network not already taken into account in the standing charge – principally that proportion of the transmission network used for the telephone service, trunk/toll switching centres, and the non-user-related element of local switching centres. In addition to these ongoing standing and call charges, one-off payments are required to cover the cost of, for example, initial connection to the network.

Trunk calls make use of large, efficient switching centres and traffic routes for the major part of their routing across the telephone network; hence their cost to the network operator per unit of distance is relatively low. Local calls, on the other hand, are confined to the smaller, less efficient switching centres and traffic routes and, in addition, the expensive local-line circuits contribute a much more significant percentage to the cost of local calls than trunk/toll calls; this results in a much higher cost to the network operator per unit of distance. It is to be expected, therefore, that trunk/toll and local calls will be charged for at differing rates.

The British Telecom approach to the structuring of call-charge rates, now also adopted by Mercury Communications for their local, trunk and international services, is based on the original division of the UK telephone network (at that time controlled by the Post Office) into so-called charging-group areas. Each charging group typically covers an area some 15–20 km in diameter, comprising a number of local switching-centre catchment areas (Fig. 10.2), and is allocated a specific STD (subscriber trunk dialling) code for incoming access. Calls are charged for on a time and distance basis, providing a very flexible tariff structure in which the charges per unit of time and the distance ranges can be varied independently. Any call within the caller's 'home' charging group and its immediately adjacent charging groups is classed as a local call, irrespective of the distance involved. Calls beyond this distance are classed as trunk, differing rates being charged for distances up to or more than 56 km (35 miles). Calls to the Irish Republic from Great Britain, and to mobile telephones irrespective of distance, are also classed as trunk and charged at a still higher rate. International calls may pass over the cable, radio or satellite paths of several different carrier organisations *en route*, requiring the revenue from call charges to be apportioned between all the organisations concerned on each particular route; calls are charged for on the basis of a much larger number of charging bands which reflect this requirement as well as the considerable variation in distances involved.

The normal type of tariff structure described above is based on the concept that all call charges are paid by the caller. There are special services, however, which allow calls to be made nationally to particular businesses or so-called information providers (a) free of charge to the caller (numbers beginning with 0800), (b) at the price of a local call (numbers beginning with 0345), or (c) at a fixed trunk-call distance rate (numbers beginning with 0055, 0066 and, most expensive 0077 or 0898). Any deficit between the price charged to callers and the true cost of the calls is effectively made up by a service charge levied on the called customer.

If time is not a factor in the charging process, a call may be recorded as

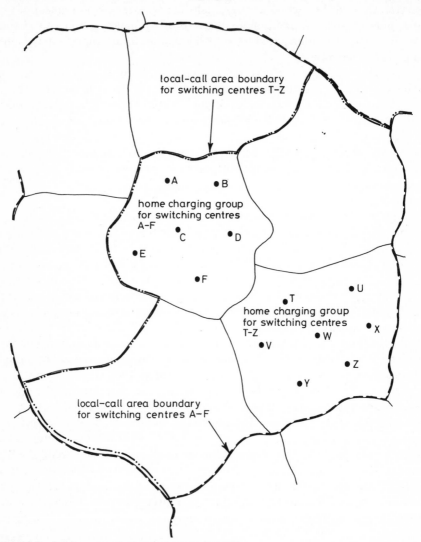

Fig. 10.2 *Public-network charging-group concept.*

successful either when the called user answers or when either party clears down. When calls are charged for on an elapsed-time basis, however, timing for charging purposes commences when the called user answers. For PBX users this only applies when *direct dialling in* (DDI) to extensions is possible; otherwise charging commences when the PBX operator answers, at increased cost to the caller. The timing process ceases when either the calling or called party clears down. However, there is a significant difference between the actual charging arrangements adopted by the two competing network operators in the UK. The Mercury Communications call-charge rates are quoted on a per-minute basis; call durations are measured precisely and the cost per call is calculated to the nearest tenth of a penny, subject to a low minimum charge for trunk calls. The British Telecom call-charge rates are based on a common unit fee, the time allowed per unit differing according to the charge rate – from a few seconds for international calls to several minutes for cheap-rate local calls; the unit fee is, therefore, the minimum charge for any type of call. Each call is metered periodically at fixed intervals determined by the charge rate applied, the number of chargeable units per call being effectively rounded to the next whole number; the effect of this rounding process on costs is of little significance for longer-distance trunk and international calls, however, because of the very short time allowed per unit.

The present British Telecom tariff structure for direct-dialled calls evolved at a time when a large business market used the financially limited capacity of the UK telephone network to the full during working hours, but the network was under-utilised at off-peak periods. It was, therefore, a sound economic policy, and one widely followed throughout the world, to charge less for off-peak calls, and also to attract new residential users by introducing an element of cross-subsidisation, in effect transferring costs from the standing charge and local-call rates to the trunk-call rates. It is the regulation of competition embodied in the Telecommunications Act of 1984 which has caused BT to progressively elimi-nate this transfer of costs in their charges, rather than the introduction of competition or the privatisation of BT *per se*. The licences issued to network operators under this Act require that all cross-subsidisation should be elimi-nated within a defined period, the intention being to prevent the established network operatior from using unfairly cross-subsidised tariffs to the disadvan-tage of a new network operator. This requirement is enforcing a relative decrease in real terms in British Telecom charges for trunk calls, and compensat-ory increases in standing and local-call charges. However, it has to be accepted that because the forces of competition have most impact on the business market – the principal users of trunk services, which have been the first to benefit from competition in the UK – the pressure on BT to reduce its tariffs in this area would inevitably have had an initially adverse effect on other charges.

The requirement that all customers on one network must have access to all customers on the other network adds a new complication to the situation where dual national networks are planned to cover the same geographical area. In the UK, the revenue from an inter-network call is apportioned between the network operators on an agreed basis. Calls are charged for in accordance with the tariff structure of the originating network and billed in the normal way by the company operating that network. Special additional metering equipment at the network interfaces enables the appropriate revenue transfers to be calculated. In most other countries where competition exists, the provision of trunk/toll service is independent of local service and in many cases each local-service provider serves a separate customer area. In the USA, generally regarded as the forerunner in competitive service, a number of long-distance carriers offer customers a choice of toll services. Most of these toll services do not give full national coverage, some being confined to one or more heavily used routes between major cities; hence they are not all available to every user. Users can elect to become a customer of as many of the available long-distance carriers as they wish, in addition to the local-service provider, and are billed separately by each company. A customer wishing to make a trunk/toll call has to dial or key a special code appropriate to the required long-distance service. In the UK, a user served by one network can similarly become a customer of the trunk service provided by the competing network. Users with a British Telecom local line can, for example, become customers of the Mercury trunk/international service. Mercury then provides either (a) on a per-line basis, a special telephone from which the Mercury code transmission is accomplished by the user pressing a single 'blue' button, or (b) for a multiline PBX, a wired-in 'smart box' which is programmed to route all within-Mercury-network and international calls to Mercury automatically.

British Telecom's peak, standard and cheap call-charge rates are now matched by Mercury's identically timed prime, standard and economy rates; in addition, Mercury offers reduced trunk-call rates for calls within its network, while BT has introduced reduced trunk-call rates for calls over selected routes. However, Mercury is as yet only able to provide a limited within-network local-call service in certain areas and is unable to give within-network access to all trunk destinations. For an accurate comparison of BT and Mercury call charges, it is necessary to calculate the cost of calls to each particular UK and international destination separately, taking into account the probable average call duration, the differing charging bases applied by the two network operators (including the added cost of a Mercury authorisation code and any special telephone(s) or a 'smart box'), and the charging periods during which calls are likely to be made.

Mobile radio network operators have their own scales of standing call charges which, as might be expected, are generally higher than those which apply to the landline networks. In the UK, however, calls from a landline-network telephone to a mobile telephone are charged for at the most expensive trunk-call rates, irrespective of distance, and it is much more costly to call into a mobile network than from it – a factor to bear in mind when deciding how best to use a mobile telephone.

Data, facsimile and teletex calls made over the public telephone network are indistinguishable from telephone calls as far as the call metering equipment is concerned and they are, therefore, subject to the same call charging arrangements. The tariffs and metering arrangements for messages sent over public telex networks follow a similar pattern to those of telephone networks, but data transmitted over a public packet-switched network is charged for on the basis of time and the amount of data transferred. Where access to the packet-switched network is obtained via the telephone or telex networks, normal local-call charges also apply. The tariffs are structured to encourage the use of the public packet-switched service in preference to the telephone network where data needs to be transferred quickly between users.

The tariffs charged to lease a public-network circuit are particularly relevant to the operator of a widely dispersed private network. Competition in the UK and USA has widened the choice open to the private-network operator who needs to transfer traffic between two separate locations. A circuit may be leased directly from one of two (or in the USA more) public networks, or – where there is capacity to spare on the links of a large private network – from the operator of that network. Leased private circuits are normally charged for on the basis of (a) the class of circuit, as determined by the frequency bandwidth (analogue) or transmission bit rate (digital) and any incorporated multiplexing arrangements, and (b) distance; as might be expected, the cost per kilometre decreases as distance increases. In the UK, standard tariff rates are generally published for 3 kHz analogue or 64 kbit/s digital circuits and circuits designed for specific data bit rates, but wideband requirements are normally costed individually. Since the Mercury network is entirely digital, analogue leased circuits cannot be provided, whereas BT is still able to provide cheaper analogue as well as digital capacity. The tariff structures and rates offered by each network must be carefully compared, not only with each other but also with the cost of other methods of transferring the required traffic, as discussed in Chapter 11.

Chapter 11

Finding the right system/network solution

11.1 Establishing the requirement

A number of different solutions may meet the operational and technical needs of a telecommunications requirement with a sufficient degree of success to warrant consideration. Among these, the 'right' solution is the one that offers the most effective answer to that communications need at the lowest cost. Choosing the right solution involves the operational, technical and financial evaluation of each alternative. The busy manager who wishes to update a private telecommunications facility may decide to place the entire responsibility for finding the right solution upon a firm of consultants or even one supplier, but this is not necessarily the most cost-effective course of action. The recommended solution could prove to be a less-than-optimum compromise, especially if the choice has been restricted to the products of one manufacturer. Even a busy manager, however, can assume at least some of this responsibility personally. At the very least, personal involvement will reduce the amount and cost of expert advice needed and help to ensure a satisfactory outcome to the project.

The manager's first task, whether or not outside technical advice is to be sought, is to establish the type and extent of communications facilities needed, and to document this information as a list of requirements. Clear specification is vital, otherwise an inappropriate solution may result. An inadequate list of requirements can also make it difficult, if not impossible, to obtain redress from a consultant or supplier if the recommended equipment subsequently proves to be unsuitable. Any list of requirements, whether intended for use personally, by a consultant, or by a potential supplier, should ideally include at least the following basic information:

(a) The method(s) of communication and, where appropriate, computing facilities required.
(b) The initial number of users, their location, and the growth expected over, typically, a four- to ten-year period.
(c) The maximum average traffic flow and, for telephone users, calling-rate

demand to be catered for per user, forecast on an annual basis throughout this period for each method of communication and computing facility separately (where some users will not require to use all methods and facilities, their distribution will need to be defined).

(d) Any performance targets to be met (Sections 10.2 and 10.3).

(e) Any special facilities or features required.

A manager may wish to seek the advice of a consultant or supplier in respect of some or all of this information, especially (c) and (d), but should ensure that the finally agreed requirement is formally documented before work on the project commences.

11.2 Narrowing the choice

Once the list of requirements has been completed, the next task is to narrow the choice of system or network concepts to be considered, by eliminating those which are clearly inappropriate for the communications need envisaged. There is unlikely to be a need for this elimination stage when the requirement is for a straightforward single-site telephone facility with access to the public network, since a PBX is the obvious solution. It may similarly be obvious that a localised data-communication requirement can best be met by a simple unswitched network or LAN, depending on the extent of communication needed (Section 5.3). In either case, the manager may wish to invite tenders directly from advertised suppliers of the required product on the basis of the prepared list of requirements. The more complex a project is, however, the more necessary it is to carry out a preliminary design study; this requires a detailed technical understanding of the conceptual possibilities. As the number of users and the distances over which they need to communicate increase, more and more options need to be considered. A cost-effective solution within the compass of a single supplier becomes increasingly difficult to achieve. Data communication in particular, especially where there is a requirement for handling telephone traffic as well, requires a detailed knowledge of the range of equipment available and the interworking problems involved, both to ensure satisfactory performance and to enable advantage to be taken of the latest technology in new systems coming on to the market. Nevertheless, the information given in earlier chapters should enable even non-technical managers at least to question and be a party to most of the decisions that have to be made.

The next task is to identify which manufacturers produce equipment that enables the selected system or network concept(s) to be realised in practice. Each manufacturer's version of a system or network concept is different; each product is individually designed to use different components and software, and differing labour and materials costs are involved in the manufacturing process. It is important to remember that the overall performance of a *system*, whether it is a self-contained PBX or LAN or a network component for switching or transmission purposes, is largely fixed by its design. A system design which ensures the efficient transfer of information between users will contribute to lower costs overall, irrespective of dimensioning/sizing considerations. Other important aspects of performance are power consumption, reliability and ease of maintenance, which affect the ongoing running cost. Equipment modularity must also be considered because it affects both the initial system cost and the cost of adding more capacity or extra facilities in the future. These and many other factors contribute to the ultimate product capability, selling price and running cost. Low initial cost does not necessarily imply an inefficient product, nor high cost a reliable one. When selecting a system, therefore, it is vital to consider both *cost* and *performance capability* in relation to the particular task it will be required to perform.

A *network* is a flexible arrangement of individual network components; once these have been carefully selected on the basis of performance and minimum cost, they can be configured and dimensioned/sized in a variety of ways. Maximum traffic-carrying efficiency is achieved by a combination of appropriate dimensioning/sizing and the selection of the optimum solution for each of a number of separate networking choices.

The product information supplied by a manufacturer is rarely sufficiently comprehensive to provide answers to all the questions that might arise. Carefully considered and probing questions to the supplier are, therefore, an essential means of determining the acceptability of a product for a particular requirement, even when the questioner has no prior knowledge of that product. To be effective, however, the questions need to be based on a sound understanding of the concept involved. An experienced independent consultant is likely to have had feedback on the operational performance of a wide range of products, and can quickly establish the relevance of competing manufacturers' claims to the particular use envisaged. A manager can and should be prepared to probe these claims in the absence of unbiased technical advice but, without expert knowledge, may be unable to establish which are the most relevant to the intended product use. Nevertheless, differences between the claims made by competing suppliers can always be highlighted and questioned to good effect.

Performance and cost factors are equally important where some private-network traffic has to be routed over a public telecommunications network, either because there is no other means of transferring this traffic to and from particular users or because it is more cost-effective to do so. Where a PBX is to be linked to a public network the only question that arises is to which network, and that only where competition exists. However, a wider range of options becomes available whenever traffic has to be conveyed between two or more geographically separated elements of a private communications facility, and the choices to be made are linked to the overall question of network design. A decision then has to be made between some or all of the following possibilities:

(a) Purchasing or leasing an independent transmission system such as a microwave radio link.
(b) Leasing a public-network line for the transmission of some (in conjunction with (d)) or all of the traffic.
(c) As an alternative to (b), leasing a line on a network operated by a large business concern for its own purposes, where spare capacity can be made available between the required locations.
(d) Transmitting some (in conjunction with (b) or (c)) or all of the traffic over a switched public network.
(e) Leasing a link via a satellite communications system.

More than one potential carrier may be available under some of these headings and a detailed knowledge of the market is essential if all possibilities are to be taken into account. Other relevant factors to be considered in addition to cost include the security of information passed over a line, the quality and reliability of the service offered, and the availability of alternative facilities in the event of breakdown.

The most commonly encountered situation arises where a choice has to be made between leasing one or more private circuits from a public-network operator, sending traffic directly over a public switched network, or using a form of alternative routing to share the traffic between them. Another important consideration, particularly for larger private telephone networks, is the greater traffic-carrying efficiency of large groups of circuits; it is possible to configure and dimension a network to take advantage of both this and the distance-related tariff structure, to minimise the overall cost of circuit leasing.

Where data transmission is concerned, any savings to be achieved by optimising transmission speed should be considered in relation to the alternative carriers' circuit-leasing costs for various transmission speeds. Although digital circuits are more expensive than analogue circuits, the latter require the use of modems, the cost of which must be taken into account. Leasing charges and modem costs generally increase with transmission speed; hence there is no point in using greater speeds than absolutely necessary. It is useful to remember that a good typist can produce data at about 5 characters per second, which equates to no more than 100 bit/s, and the average person cannot read text appearing on a screen at more than 300 bit/s. Transmission speeds of this order are, therefore, quite adequate for many data transmission applications. If, on the other hand, there is a need frequently to transfer a full screen of data (typically around 2000 characters) or the contents of a floppy disc (say about 350 000 characters), higher transmission speeds become economic because of the savings in time that can be achieved, as indicated in the following table:

Transmission speed (bit/s)	Approx. time to transmit	
	Screen	*Floppy disc*
300	1 min	3 h
1200	15 s	45 min
2400	7 s	21 min
4800	4 s	11 min
9600	2 s	6 min

11.3 Finding the most cost-effective solution

The process of finding the most cost-effective solution for any private telecommunications facility, however complex, requires a number of separate areas of network choice to be first identified and then evaluated. The most important

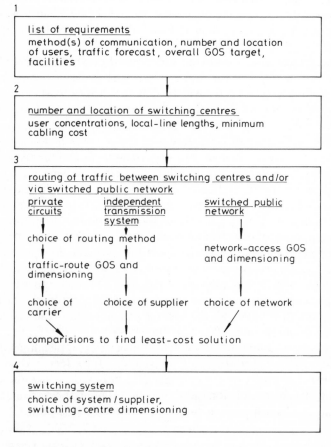

Fig. 11.1 *Private telephone network options.*

factors to be taken into account are restated in Figs. 11.1 (telephone networks) and 11.2 (data networks). Once the overall network concept has been finalised in outline, it is necessary first to identify those aspects and components where an option exists for alternative solutions, and then to establish any interrelationships between them. The possible alternatives for each network choice are

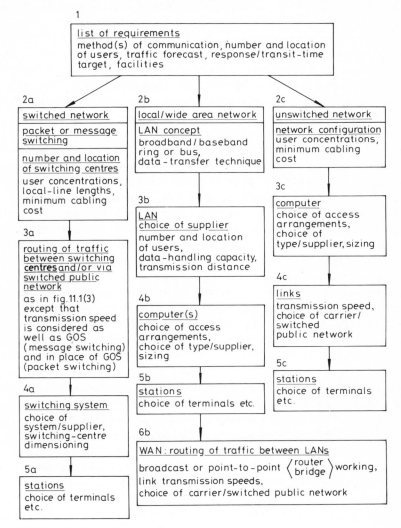

Fig. 11.2 *Private data network options.*

reduced, by a process of elimination, to a shortlist of proprietary/practical options which are fully, or to an acceptable extent, capable of meeting the requirement and are, therefore, appropriate to be costed. Some decisions, such as which telephone switching system should be used, can be taken in isolation;

others, for example the means of interconnection to be employed for a data network, are inextricably linked to other network design questions. Network design must, therefore, proceed in concert with these comparisons, and at least broad-brush dimensioning/sizing is necessary in order to determine the costs involved. Where it is impracticable to put a realistic price on some of the factors to be taken into account, a subjective judgment has to be made on their importance to the business relative to cost. The best overall solution could be determined by incorporating the resulting series of comparisons of costs and other factors, each relating to a different question and concerning one aspect of the total network requirement, in a single comprehensive study. In practice, however, it is often easier to undertake each independent comparison separately, progressing towards an optimum overall solution on a step-by-step basis.

Where the costs for each alternative solution can be reduced to an annually repeated relationship which is not expected to change significantly for the foreseeable future – as, for example, where the expected cost of public-network calls is to be compared with the cost of leasing private circuits – those for a single year may be compared directly. More usually, however, the overall cost comprises (a) one-off capital payments, both initial and at differing times in the future for each option, and (b) various periodic charges for equipment and/or circuit leasing, public-network calls, etc., together with the ongoing running costs, all of which may have a changing value with time which can be expressed on an annual basis. For a fair and accurate comparison between options, it is then necessary to take into account the so-called *time value of money*, whereby money required for expenditure at some time in the future can be considered to be invested to earn interest, requiring a lesser present-day sum to be set aside than if the expenditure were required immediately. 'Future' money is expressed in terms of its *present value,* calculated in practice by multiplying each sum by a present-value factor which takes into account an appropriate discount rate for the required number of years. This is the basis of the *discounted cash flow* (DCF) technique, widely used in telecommunications and computing practice. In a DCF study, the series of comparisons between different purchase/running costs, leasing terms, public-network tariffs, etc. should each cover the period for which forecasts have been prepared, usually at least four and no more than ten years.

A step-by-step selection of the most cost-effective alternative for each network choice, together with the intelligent application of performance targets and dimensioning/sizing rules, will result in a comprehensive least-cost solution for any telecommunications facility. However, the selection process described can be both time consuming and costly if every possible alternative is to be considered, especially for large, complex networks. Ideally, the costs of undertaking a full-scale evaluation should be weighed against the savings likely to be achieved – probably a very small percentage of the cost of the facility to be provided – and it would be logical to match the thoroughness of each evaluation to the needs of the business concerned. In practice, however, a much more pragmatic approach can be expected. Evaluations are likely to follow a standard pattern; only a few selected network choices will be subjected to a detailed comparison, the majority being decided on the basis of past experience, company or personal preference for particular products, or some other less tangible reason. It is unlikely that a network based on a fully comprehensive evaluation would perform significantly better or cost significantly less than one where the unevaluated decisions have been based on intelligently applied experience. The same cannot always be said of networks designed on the basis of less informed decisions.

The success of an operational telecommunications or computing network can only be judged in very broad terms in the absence of detailed measurements: does it work satisfactorily or is it obviously overloaded? In the vast majority of cases, networks do meet the needs they have been *designed* to meet quite satisfactorily, but there is no doubt that sometimes the design requirements do not match the business needs or success is achieved at an unnecessarily high cost. If a network is found to be overloaded, or inadequate in some other respect for the required task, it may prove to be both difficult for the business to prove responsibility and costly to remedy the situation. By becoming more involved in the provision of telecommunications and computing facilities, a manager can ensure that the particular needs of the business are not only fully understood, but also met at the lowest possible cost.

11.4 Cost-comparison examples

The examples in this section have been chosen to illustrate important points made in the book. All are worked in full to demonstrate the principles involved, but short cuts will become apparent once the methods are understood. In the interest of simplicity, the examples do not necessarily cover all the options that would need to be considered in practice; attention is drawn to omitted options in the text of each example. The system and network equipment featured, and the costs applied to them, are purely hypothetical and in no way intended to be representative of particular proprietary products; tariffs and leasing charges, although broadly based on real examples, should similarly be regarded as hypothetical. The results stem from a particular combination of costs, traffic levels, distances, etc. and should *not* be seen as inferring a standard pattern for practical decision making. Unless there are any other overriding factors which require a specific course of action to be followed, each case where a selection has to be made between competing carriers, systems and network design options should be individually costed.

The following basic rules have been observed in applying the discounted cash flow (DCF) cost-comparison method:

(a) The year of initial purchase is designated as year 0, and subsequent years as year 1, year 2, etc.

(b) Purchase costs are taken into account at the beginning of the year in which they are incurred and running costs at the end of the year to which they apply.
(c) For simplicity, a common discounting rate of 10% has been applied to all examples. The appropriate present value (PV) factors are: year 1 = 0.9091; year 2 = 0.8264; year 3 = 0.7513; year 4 = 0.6830.

Example 1
A small company rents four exchange lines from the established public telephone network A. A choice has to be made between continuing to use this network for trunk calls, principally to two other cities X and Y, and for international calls to the USA, or using a newly established network B for this purpose. Whatever the decision taken, network A will continue to be essential for local traffic.

Traffic and costing information: The estimated annual total of calls is as follows:

City X: 1500 at peak rate, 1000 at standard rate
City Y: 900 at peak rate, 500 at standard rate
USA: 150 at peak rate,   60 at standard rate

The estimated average call duration = 2.5 minutes.

Call charges made by the two network operators are:

Network A
City X: 4.5p per 15s peak, 4.5p per 20s standard
City Y: 4.5p per 20s peak, 4.5p per 30s standard
USA: 4.5p per 4s peak, 4.5p per 4.5s standard

City Y is via a 'low-cost' route: hence charging is lower.

Network B
City X: 11.5p per min peak, 9.5p per min standard
City Y: 12.5p per min peak, 10.5p per min standard
USA: 60.5p per min peak, 55.5p per min standard

City Y is not within network B: hence charging rate is higher.

Additional one-off costs for special telephones and an authorisation code are incurred by using network B, totalling £250.

Cost comparison: Note that time units (TUs: calculations in square brackets [ ]) are rounded to the next whole number before being multiplied.

(a) Annual cost of calls to city X:

Network A

Peak: TUs = [2.5 × 60/15] = 10:   × 0.045 × 1500 = £ 675
Standard: TUs = [2.5 × 60/20] = 8:   × 0.045 × 1000 = £ 360
                                                                Total = £1035

Network B

Peak: 0.115 × 2.5 × 1500 = £431
Standard: 0.095 × 2.5 × 1000 = £238
                                Total = £669

(b) Annual cost of calls to city Y:

Network A

Peak: TUs = [2.5 × 60/20] = 8:   × 0.045 × 800 = £288
Standard: TUs = [2.5 × 60/30] = 5:   × 0.045 × 500 = £113
                                                            Total = £401

Network B

Peak: 0.125 × 2.5 × 800 = £250
Standard: 0.105 × 2.5 × 500 = £131
                                Total = £381

(c) Annual cost of calls to USA:

Network A

Peak: TUs = [2.5 × 60/4.0] = 38:   × 0.045 × 150 = £257
Standard: TUs = [2.5 × 60/4.5] = 34:   × 0.045 × 60 = £ 92
                                                              Total = £349

Network B

Peak: 0.605 × 2.5 × 150 = £227
Standard: 0.555 × 2.5 × 60 = £ 83
                                Total = £310

Summary of results and conclusions:

Network A
Annual cost of calls via network A = £1035 + 401 + 349 = £1785.

Network B
Annual cost of calls via network B = £ 669 + 381 + 310 = £1360.
Additional costs for first year only are £250.
Hence total for first year is £1610, and annual cost for subsequent years is £1360.

From these results, it is clearly cheaper for the company to route trunk and international traffic via network B. With smaller numbers of calls required, however, there will be a break-even point below which it will be cheaper to continue to use network A for this purpose.

Example 2
A company requires telephone communication between three widely separated PBXs, designated for the purpose of this example as A, B and C. A choice must be made between sending the inter-PBX traffic via a public telephone network, over leased private circuits, or sharing it between them.

Traffic, dimensioning and costing information:
The distances between the PBXs are:

A to B = 130 km
A to C = 200 km
B to C = 120 km

The average number of calls forecast to be required between the PBXs during the busy hour is as follows:

A and B: 15
A and C: 10
B and C: 30

The total traffic during a working day is expected to be approximately five times that during the busy hour; the company operates an 8 hour working day for 253 days per year.

The overall GOS target has been set at 0.05 (this is within the range expected of a typical public-network connection, depending on the number of links involved). Because some destination telephones will inevitably be engaged or remain unanswered during the busy hour, unsuccessful call attempts in excess of the numbers of successful calls forecast will also engage the interconnecting leased circuits and/or public-network access (local) circuits; a proportion of 7 successful to 3 unsuccessful call attempts is assumed. Although successful calls only are charged for over the public network, any leased private circuits and public-network access circuits need to be dimensioned to take this additional unsuccessful-call-attempt traffic into account. The average call holding time during which a circuit is engaged for a successful call is predicted to be 200 s, compared with 40 s for an unsuccessful call attempt.

A preliminary assessment of the charges made by competing network operators shows that the following charges are the cheapest available:

Cost of a 200s call between A and B and between B and C is £0.60 peak (a.m.) and £0.45 standard (p.m.).
Cost of a 200s call between A and C is £0.45 peak (a.m.) and £0.30 standard (p.m.).
Annual cost of leasing one circuit = £1100 + £10 per km over 15 km.
Annual cost of one additional public-network access circuit = £100.

Cost of sending all traffic via leased private circuits:
Total time for which circuits are expected to be engaged during the busy hour:

Route A–B: $15 \times 200\,s + (3/7) \times 15 \times 40\,s = 3257\,s$
Route A–C: $10 \times 200\,s + (3/7) \times 10 \times 40\,s = 2171\,s$
Route B–C: $30 \times 200\,s + (3/7) \times 30 \times 40\,s = 6514\,s$

Traffic flow in erlangs = time in seconds/3600, giving:

Route A–B: approx. 0.9 E
Route A–C: approx. 0.6 E
Route B–C: approx. 1.8 E

Annual leasing charge per circuit:

Route A–B: $[£1100 + (105 \times £10)] = £2150$
Route A–C: $[£1100 + (185 \times £10)] = £2950$
Route B–C: $[£1100 + (115 \times £10)] = £2250$

The GOS for dimensioning is to be 0.05. From the traffic table at Appendix A the number of circuits required on each route, and hence the annual cost, is as follows:

Route A–B: 3 circuits at £2150 = £ 6 450
Route A–C: 3 circuits at £2950 = £ 8 850
Route B–C: 5 circuits at £2250 = £11 250

Cost of sending all traffic via the public network: In the absence of a detailed daily traffic profile, an assumption must be made concerning the proportion of calls made during the morning peak-rate and afternoon standard-rate periods. Typically, calls are divided almost equally between these two periods; on this assumption, the average cost per call will be (0.60 + 0.45)/2 = £0.525 for routes A–B and B–C, and (0.45 + 0.30)/2 = £0.375 for route A–C.

Therefore, by multiplying the busy-hour call figure by $5 \times 253 \times 0.525 = 664$

for routes A–B and B–C, and by 5 × 253 × 0.375 = 474 for route A–C, the following annual costs are obtained:

Route A–B: 15 × 664 = £ 9 960
Route A–C: 10 × 474 = £ 4 740
Route B–C: 30 × 664 = £19 920

In addition, the number of public-network access circuits will need to be increased at each affected site, the worst-case situation being where existing lines are already loaded to capacity. Assuming circuit quantities as for leased circuits, the maximum additional annual cost will be for three extra lines at each of A and B for route A–B, namely £600; three extra lines at each of A and C for route A–C, namely £600; and five extra lines at each of B and C for route B–C, namely £1000. These figures are acceptable for cost-comparison purposes but in practice, if two or all of these traffic streams are to be routed via the public network, the overall number of lines required at each site and hence the extra cost involved will be less than the arithmetic sum because the aggregate lines will constitute a larger, more efficient traffic-carrying group.

Hence total costs are as follows:

Route A–B: £ 9 960 + £ 600 = £10 560
Route A–C: £ 4 740 + £ 600 = £ 5 340
Route B–C: £19 920 + £1000 = £20 920

Cost of sending traffic via leased private circuits with overflow to the public network: This option is only possible where the PBXs are equipped to provide the appropriate fixed alternative routing facility, for which it is assumed there is no extra charge; if, in practice, extra cost is incurred, this would need to be taken into account in the final analysis.

Following the recommended procedure from Section 8.1 and consulting the table at Appendix B, the revised leased-circuit quantities and annual cost will be:

Route A–B: 2 circuits at £2150 = £4300
Route A–C: 2 circuits at £2950 = £5900
Route B–C: 4 circuits at £2250 = £9000

From the traffic table at Appendix A, two circuits can carry 0.38 E at a GOS of 0.05 and four circuits can carry 1.53 E. Hence the overflow traffic to the public network from each route can be assumed to be as follows:

Route A–B: 0.9 − 0.38 = 0.52 E
Route A–C: 0.6 − 0.38 = 0.22 E
Route B–C: 1.8 − 1.53 = 0.27 E

These overflow traffic flows will, from the traffic table at Appendix A for a GOS of 0.05, require three, two and two additional public-network access lines at each site, giving rise to extra costs of £600, £400 and £400 respectively.

Since the majority of traffic overflows are bound to occur during the busy hour, the assumption made in the previous section concerning the spread of calls throughout a working day is no longer valid. In the absence of a detailed traffic profile, a more realistic assumption is that two-thirds of the overflowing calls will occur during the busy hour, with three-quarters being charged for at the morning peak rate.

The changed assumption means that the total daily overflow traffic will be only 1.5 times the busy-hour figure, and the average cost per call is $[(3 \times 0.60) + (1 \times 0.45)]/4 = £0.5625$ for routes A–B and B–C, and $[(3 \times 0.45) + (1 \times 0.30)]/4 = £0.4125$ for route A–C. The annual cost factor then becomes $1.5 \times 253 \times 0.5625 = 213$ for routes A–B and B–C, and $1.5 \times 253 \times 0.4125 = 157$ for route A–C. The annual costs are then as follows:

Route A–B: $(0.52/0.90) \times 15 \times 213 = £1846$
Route A–C: $(0.22/0.60) \times 10 \times 157 = £\ 576$
Route B–C: $(0.27/1.80) \times 30 \times 213 = £\ 959$

Hence total costs are as follows:

Route A–B: £4300 + £1846 + £600 = £ 6746
Route A–C: £5900 + £ 781 + £400 = £ 7081
Route B–C: £9000 + £ 959 + £400 = £10359

Summary of results and conclusions: The annual costs for each route for the three traffic options are as follows:

Route	*Public network*	*Leased circuits without overflow*	*High-usage leased circuits with overflow to PN*
A–B	£10 560	£ 6 450	£ 6 746
A–C	£ 5 340	£ 8 850	£ 7 081
B–C	£20 920	£11 250	*£10 359*

The results show that the cheapest options for each route are as follows:

Route A–B: leased circuits without overflow
Route A–C: public network
Route B–C: high-usage leased circuits with overflow to public network.

Example 3

A company wishes to choose between two PBXs, designated here as X and Y, from competing manufacturers. The PBXs are broadly comparable in claimed performance and reliability and the facilities offered; therefore the choice has to be made on the basis of cost alone.

Traffic and costing information: The requirements to be met are:

	At opening	Year 1	Year 2	Year 3
No. extensions	420	500	520	560
No. PN + tie lines	50	60	63	70

Average busy-hour calling rate per extension: 0.40 throughout.

Average busy-hour incoming calling rate per PN/tie line: 0.50 throughout.

These figures necessitate a maximum call-attempt capacity of (560×0.40) + $(70 \times 0.5) = 259$ call attempts per hour. In order to simplify the comparison, each PBX is assumed to be functionally similar and in accordance with the typical configuration described in Section 5.2 and illustrated in Fig. 5.3. The capacity and cost figures are as follows:

	PBX X	PBX Y
(a) Equipment module capacity and cost:		
Control unit:		
Busy-hour call attempt capacity	350	400
Cost (including associated eqpt)	£14 120	£14 760
Combined line-termination and switch units:		
Basic provision for extensions	216	128
PN and tie lines	32	16
Cost of basic provision	£26 720	£25 216
Add-on units, per unit: extensions	108	128
PN/tie lines	16	16
Cost per add-on unit	£15 100	£15 880
Maximum no. of add-on units possible	4	5
Peripherals units:		
Average cost per 108 working extensions	£ 5 480	–
64 working extensions	–	£ 3 040
Operator's console:		
Cost	£ 1 440	£ 1 380
(b) Average running cost (maintenance plus power):		
Per annum basic	£12 000	£10 000
Plus per add-on unit	£ 1 800	£ 1 300

Costs of PBX X:
Initial purchase:

Control unit	£14 120
Line/switch units: basic	£26 720
add-on × 2	£30 200
Peripherals units × 4	£21 920
Operator's console	£ 1 440
Total cost	£94 400

Additional purchase, year 1:

Line/switch unit: add-on × 1	£15 100
Peripherals unit × 1	£ 5 480
Total cost	£20 580

Additional purchase, year 3:

Line/switch unit: add-on × 1	£15 100
Peripherals unit × 1	£ 5 480
Total cost	£20 580

Running costs:

End of year 0	£12 000 + (2 × £1800) = £15 600
End of year 1	£12 000 + (3 × £1800) = £17 400
End of year 2	£12 000 + (3 × (1800) = £17 400
End of year 3	£12 000 + (4 × £1800) = £19 200

Costs of PBX Y:
Initial purchase:

Control unit	£ 14 760
Line/switch units: basic	£ 25 216
add-on × 3	£ 47 640
Peripherals units × 7	£ 21 280
Operator's console	£ 1 380
Total cost	£110 276

Additional purchase, year 1:

Peripherals unit × 1	£ 3 040

Additional purchase, year 2:

Line/switch units: add-on × 1	£15 880
Peripherals unit × 1	£ 3 040
Total cost	£18 920

Running costs:

End of year 0	£10 000 + (3 × £1300) = £13 900
End of year 1	£10 000 + (3 × £1300) = £13 900
End of year 2	£10 000 + (4 × £1300) = £15 200
End of year 3	£10 000 + (4 × £1300) = £15 200

DCF study:
(a) Timing of costs:

Year	0	1	2	3	4
PBX X					
Purchase	£ 94 400	£20 580		£20 580	
Running		£15 600	£17 400	£17 400	£19 200
PBX Y					
Purchase	£110 276	£ 3 040	£18 920		
Running		£13 900	£13 900	£15 200	£15 200

(b) Comparison of costs:

Year	Total cost (£) PBX X	PBX Y	PV factor	Present value (£) PBX X	PBX Y
0	94 400	110 276	1.0000	94 400	110 276
1	36 180	16 940	0.9091	32 891	15 400
2	17 400	32 820	0.8264	14 379	27 122
3	37 980	15 200	0.7513	28 534	11 420
4	19 200	15 200	0.6830	13 114	10 382
Total present value				183 318	174 600

Summary of results and conclusions: In terms of present value over four years, PBX Y, although initially more expensive, is cheaper than PBX X by almost £9000. An added factor in favour of PBX Y is that its maximum capacity is greater than that of PBX X should continuing growth be required.

Example 4
A company, having decided to purchase a PBX to serve a new manufacturing site S, wishes to link this into an existing network serving three existing sites P, Q and R. Site Q is equipped with tandem switching facilities, and tie lines link this site to each of P and R. It is proposed (a) to rationalise the provision of tie lines throughout the network by applying an overall GOS standard of 0.1, and (b) to determine the most economic means of linking the new PBX to the existing sites.

Traffic and costing information: The existing tie lines and distances between the sites are as follows:

P–Q:	25 circuits	75 km
Q–R:	25 circuits	85 km
S–P:		45 km
S–Q:		55 km
S–R:		120 km

The traffic forecast for years 3 and 5, assuming the date at which PBX S is to be brought into service is year 0, is as follows:

	Year 3	Year 5
P–Q	17 E	19 E
P–R	5 E	6 E
Q–R	15 E	17 E
S–P	3 E	4 E
S–Q	6 E	7 E
S–R	4 E	5 E

The following charges are assumed for the purpose of this example:

Annual cost of leasing one circuit = £1200 + £10 per km over 15 km

The resulting costs for the distances involved are:

Tie lines	Distance	Cost
S–P	45 km	£1500
S–Q	55 km	£1600
P–Q	75 km	£1800
Q–R	85 km	£1900
S–R	120 km	£2250

The switching centre at site Q is already equipped to switch up to 10 E of tandem traffic. The cost of additional tandem switching capacity is as follows:

Equipment purchase cost per 10 E of tandem traffic = £640
Running cost per annum per 10 E of tandem traffic = £40

Possible alternative solutions:

(a) Separate leased circuits between S and P, S and Q, S and R.
(b) Leased circuits between S and Q only; traffic between S and P, S and R tandem switched at Q.
(c) Leased circuits between S and Q, S and P; traffic between S and R tandem switched at Q.
(d) Leased circuits between S and Q, S and R; traffic between S and P tandem switched at Q.

In practice, the possibility and cost of leasing circuits from different carriers, using high-usage leased circuits with overflow to a public network and routing some traffic entirely over a public network, might also need to be considered.

Costs for alternative (a): Additional requirements   The number of circuits required for each route is determined from the traffic table at Appendix A, for the appropriate GOS. Tandem switching at Q will continue to be used for traffic between P and R only; routes P–Q and Q–R, both of which carry some tandem-switched traffic, must therefore each be dimensioned to a GOS of 0.05 to achieve the overall GOS target of 0.1. The three new routes S–P, S–Q and S–R carry direct traffic only and must be dimensioned to a GOS of 0.1. The tandem switching requirement at Q only amounts to 6 E at year 5; hence no additional equipment is needed.

Route	Traffic (E)		GOS	No. of circuits	
	Year 3	Year 5		Year 3	Year 5
P–Q	17 + 5	19 + 6	0.05	28	31
Q–R	15 + 5	17 + 6	0.05	25	29
S–P	3	4	0.1	6	7
S–Q	6	7	0.1	9	10
S–R	4	5	0.1	7	8

Additional leased private circuit cost per annum:

Route	Cost per circuit	Added circuits and total cost	
		First 3 years	Next 3 years
P–Q	£1800	3 ccts = £ 5 400	6 ccts = £10 800
Q–R	£1900	none	4 ccts = £ 7 600
S–P	£1500	6 ccts = £ 9 000	7 ccts = £10 500
S–Q	£1600	9 ccts = £14 400	10 ccts = £16 000
S–R	£2250	7 ccts = £15 750	8 ccts = £18 000
		Total £44 550	£62 900

Costs for alternative (b): Additional requirements   Tandem switching will be used for traffic between P and R, S and P, and S and R; all routes (P–Q, Q–R and S–Q) will, therefore, carry some tandem-switched traffic and must be dimensioned to a GOS of 0.05. The tandem switching requirement at Q = 5 + 3 + 4 = 12 E at year 3 and 6 + 4 + 5 = 15 E at year 5; therefore an additional equipment purchase cost of £640 will be incurred at the beginning of year 0, together with an added running cost of £40 for each year of the costing period.

Route	Traffic (E) Year 3	Year 5	GOS	No. of circuits Year 3	Year 5
P–Q	17 + 5 + 3	19 + 6 + 4	0.05	31	35
Q–R	15 + 5 + 4	17 + 6 + 5	0.05	30	34
S–Q	6 + 3 + 4	7 + 4 + 5	0.05	18	21

Additional leased private circuit costs per annum:

Route	Cost per circuit	Added circuits and total cost First 3 years	Next 3 years
P–Q	£1800	6 ccts = £10 800	10 ccts = £18 000
Q–R	£1900	5 ccts = £ 9 500	9 ccts = £17 100
S–Q	£1600	18 ccts = £28 800	21 ccts = £33600
		Total £49 100	£68 700

Costs for alternative (c): Additional requirements Tandem switching will be used for traffic between P and R, and S and R; routes P–Q, Q–R and S–Q will, therefore, carry some tandem traffic and must be dimensioned to a GOS of 0.05. The remaining route S–P will carry direct traffic only and must be dimensioned to a GOS of 0.1. The tandem switching requirement at Q is 5 + 4 = 9 E at year 3 and 6 + 5 = 11 E at year 5; therefore an additional equipment purchase cost of £640 will be incurred at, say, the beginning of year 4, together with an added running cost of £40 for years 4 and 5.

Route	Traffic (E) Year 3	Year 5	GOS	No. of circuits Year 3	Year 5
P–Q	17 + 5	19 + 6	0.05	28	31
Q–R	15 + 5 + 4	17 + 6 + 5	0.05	30	34
S–P	3	4	0.01	6	7
S–Q	6 + 4	7 + 5	0.05	15	17

Additional leased private circuit costs per annum:

Route	Cost per circuit	Added circuits and total cost First 3 years	Next 3 years
P–Q	£1800	3 ccts = £ 5 400	6 ccts = £10 800
Q–R	£1900	5 ccts = £ 9 500	9 ccts = £17 100
S–P	£1500	6 ccts = £ 9 000	7 ccts = £10 500
S–Q	£1600	15 ccts = £24 000	17 ccts = £27 200
		Total £47 900	£65 600

Costs for alternative (d): Additional requirements   Tandem switching will be used for traffic between P and R, and S and P; routes P–Q, Q–R and S–Q will, therefore, carry some tandem traffic and must be dimensioned to a GOS of 0.05. The remaining route S–R will carry direct traffic only and must be dimensioned to a GOS of 0.1. The tandem switching requirement at Q only amounts to 6 + 4 = 10 E at year 5; hence no additional equipment is needed.

Route	Traffic (E) Year 3	Year 5	GOS	No. of circuits Year 3	Year 5
P–Q	17 + 5 + 3	19 + 6 + 4	0.05	31	35
Q–R	15 + 5	17 + 6	0.05	25	29
S–Q	6 + 3	7 + 4	0.05	14	16
S–R	4	5	0.1	7	8

Additional leased private circuit costs per annum:

Route	Cost per circuit	Added circuits and total cost First 3 years	Next 3 years
P–Q	£1800	6 ccts = £10 800	10 ccts = £18 000
Q–R	£1900	none	4 ccts = £ 7 600
S–Q	£1600	14 ccts = £22 400	16 ccts = £25 600
S–R	£2250	7 ccts = £15 750	8 ccts = £18 000
		Total £48 950	£69 200

Summary of results and conclusions: An examination of the total annual leasing costs for each alternative reveals that (a) is the least expensive throughout the costing period:

Alternative	First 3 years	Next 3 years
(a)	£44 550	£62 900
(b)	£49 100	£68 700
(c)	£47 900	£65 600
(d)	£48 950	£69 200

Purchase/running costs are incurred only with two of the more expensive alternatives (b) and (c), and hence do not affect the issue. By ignoring these costs, a full-scale DCF comparison has been avoided, and a decision in favour of alternative (a) can be made on the basis of the annual leasing charges alone.

Example 5
A company wishes to set up a new unswitched data communications network, which requires territorial offices X, Y and Z to be linked to a headquarters central computer over leased public-network digital circuits in the most cost-effective manner. The performance of the selected computer is such that the average response time of each link must not exceed 2.5 seconds if the company's overall response-time requirement is to be met.

Traffic and performance information:

	X	Y	Z
No. of terminals	100	65	90
Average busy-hour message pairs per terminal	25	25	25
Average message-pair duration at 1200 bit/s	80 s	80 s	80 s

Link terminating equipment is available to operate at 19.2, 48, 56 or 64 kbit/s and, on the basis of the claimed leased circuit error performance, the equipment manufacturer recommends a circuit occupancy of no more than 0.85 to ensure that the target response time of 2.5 seconds will be met.

Possible solutions: Average busy-hour data throughput:

Per terminal: $(25 \times 80/3600) \times 1200 = 667$ bit/s
Link X–HQ: 667×100 $= 66.7$ kbit/s
Link Y–HQ: 667×65 $= 43.4$ kbit/s
Link Z–HQ: 667×90 $= 60.0$ kbit/s

This requires a throughput capacity of at least the following on each link if the circuit occupancy is not to exceed 0.85:

Link X–HQ: $66.7/0.85 = 78.5$ kbit/s
Link Y–HQ: $43.4/0.85 = 51.1$ kbit/s
Link Z–HQ: $60.0/0.85 = 70.6$ kbit/s

The requirement for each link will be met by the least costly combination of link terminating equipment(s) and leased digital circuit(s), selected from the range of transmission speeds possible, which provides at least the calculated throughput capacity. For example, link X–HQ could be provided over

(a) 1×64 kbit/s and 1×19.2 kbit/s circuits, giving a throughput capacity of 83.2 kbit/s, with the 100 terminals split between them pro rata 77:23; or

(b) 2 × 48 kbit/s circuits, giving a capacity of 96 kbit/s, with 50 terminals per circuit; or

(c) 5 × 19.2 kbit/s circuits, also giving a capacity of 96 kbit/s, with 20 terminals per circuit.

Similarly, link Y–HQ could be provided over 1 × 56 kbit/s circuit or 3 × 19.2 kbit/s circuits, and link Z–HQ over 1 × 56 kbit/s and 1 × 19.2 kbit/s circuits or 4 × 19.2 kbit/s circuits.

Bibliography

BARTREE, THOMAS C. (ed.) (1985): *Data Communications Networks and Systems* (Howard W. Sams)

BEAR, D. (1980): *Telecommunications Traffic Engineering* (IEE Telecommunications Series no. 2) (Peter Peregrinus)

BINGHAM, JOHN E and DAVIES, GARTH W.P. (1977): *Planning for Data Communications* (Macmillan)

BLEAZARD, G.B. (1982): *Handbook of Data Communications* (National Computing Centre)

BREWSTER, R.L. (1986): *Telecommunications Technology* (Ellis Horwood)

CARTER, G. (1984): *Local Area Networks* (Heinemann Computers in Education Series, in conjunction with ICL)

CLARE, C.P. (1984): *A Guide to Data Communications* (Castle House)

CLARK, A.P. (1983): *Principles of Digital Data Transmission* (Pentech)

COUGHLIN, VINCE (1984): *Telecommunications Equipment Fundamentals and Network Structures* (Van Nostrand Reinhold)

DEASINGTON, R.J. (1985): *X25 Explained: Protocols for Packet Switching Networks* (Ellis Horwood)

FERRARI, DOMENICO (1978): *Computer Systems Performance Evaluation* (Prentice-Hall)

GEE, K.C.E. (1981): *Proprietary Network Architectures* (National Computing Centre)

GEE, K.C.E. (1982): *Local Area Networks* (National Computing Centre)

GRIFFITHS, J.M. (1986): *Local Telecommunications* (IEE Telecommunications Series no. 10) (Peter Peregrinus)

HALSALL, FRED (1985): *Introduction to Data Communications and Computer Networks* (Addison-Wesley)

HARPER, J.M. (1986): *Telecommunications and Computing: The Uncompleted Revolution* (Communication Educational Services)

HELD, GILBERT and SARCH, RAY (1983): *Data Communications – A Comprehensive Approach* (McGraw-Hill)

KUECKEN, JOHN A. (1983): *Talking Computers and Telecommunications* (Van Nostrand Reinhold)

LANGLEY, GRAHAM (1983): *Telecommunications Primer* (Pitman)

LITTLECHILD, S.C. (1979): *Elements of Telecommunications Economics* (IEE Telecommunications Series no. 7) (Peter Peregrinus)

MARTIN, JAMES (1969): *Telecommunications and the Computer* (Prentice-Hall)

MAYNARD, JEFF (1984): *Computer and Telecommunications Handbook* (Granada)

MORGAN, T.J. (1976) *Telecommunications Economics* (Technicopy)

NEWMAN, JOHN L. (1986); *Computer Systems: Software and Architecture* (Hutchinson)

PEARCE, J. GORDON (1981): *Telecommunications Switching* (Plenum)

SCOTT, P.R.D. (1983): *Reviewing Your Data Transmission Network* (National Computing Centre)
STAMPER, DAVID A. (1986): *Business Data Communications* (Benjamin/Cummings)
STUCK, B.W. and ARTHURS, E. (1985): *A Computer and Communications Network Performance Analysis Primer* (Prentice-Hall)

Appendix A
Full-availability traffic table for various grades of service

No. of circuits	| lost call in					
	10 (0.1)	20 (0.05)	33 (0.03)	50 (0.02)	100 (0.01)	1000 (0.001)
	E	E	E	E	E	E
1	0.1	0.05	0.03	0.02	0.01	0.001
2	0.60	0.38	0.28	0.22	0.15	0.046
3	1.27	0.90	0.72	0.60	0.45	0.19
4	2.05	1.53	1.26	1.10	0.90	0.44
5	2.9	2.2	1.9	1.7	1.4	0.8
6	3.8	3.0	2.5	2.3	1.9	1.1
7	4.7	3.7	3.3	2.9	2.5	1.6
8	5.6	4.5	4.0	3.6	3.2	2.1
9	6.5	5.4	4.7	4.3	3.8	2.6
10	7.5	6.2	5.5	5.1	4.5	3.1
11	8.5	7.1	6.3	5.8	5.2	3.6
12	9.5	8.0	7.1	6.6	5.9	4.2
13	10.5	8.8	8.0	7.4	6.6	4.8
14	11.5	9.7	8.8	8.2	7.4	5.4
15	12.5	10.6	9.7	9.0	8.1	6.1
16	13.5	11.5	10.5	9.8	8.9	6.7
17	14.5	12.5	11.4	10.7	9.6	7.4
18	15.5	13.4	12.2	11.5	10.4	8.0
19	16.6	14.3	13.1	12.3	11.2	8.7
20	17.6	15.2	14.0	13.2	12.0	9.4
21	18.7	16.2	14.9	14.0	12.8	10.1
22	19.7	17.1	15.8	14.9	13.7	10.8
23	20.7	18.1	16.7	15.7	14.5	11.5
24	21.8	19.0	17.6	16.6	15.3	12.2
25	22.8	20.0	18.5	17.5	16.1	13.0
26	23.9	20.9	19.4	18.4	16.9	13.7

No. of circuits	1 lost call in					
	10 (0.1)	20 (0.05)	33 (0.03)	50 (0.02)	100 (0.01)	1000 (0.001)
27	24.9	21.9	20.3	19.3	17.7	14.4
28	26.0	22.9	21.2	20.2	18.6	15.2
29	27.1	23.8	22.1	21.1	19.5	15.9
30	28.1	24.8	23.1	22.0	20.4	16.7
31	29.2	25.8	24.0	22.9	21.2	17.4
32	30.2	26.7	24.9	23.8	22.1	18.2
33	31.3	27.7	25.8	24.7	23.0	18.9
34	32.4	28.7	26.8	25.6	23.8	19.7
35	33.4	29.7	27.7	26.5	24.6	20.5
36	34.5	30.7	28.6	27.4	25.5	21.3
37	35.6	31.6	29.6	28.3	26.4	22.1
38	36.6	32.6	30.5	29.3	27.3	22.9
39	37.7	33.6	31.5	30.1	28.2	23.7
40	38.8	34.6	32.4	31.0	29.0	24.5
45	44.2	39.6	37.2	35.6	33.4	28.5
50	49.6	44.5	41.9	40.3	37.9	32.5
55	55.0	50.0	46.7	45.0	42.4	36.7
60	60.4	54.6	51.6	49.7	46.9	40.8
70	71.3	64.7	61.3	59.1	56.0	49.2
80	82.2	74.8	71.1	68.6	65.4	57.8
90	93.1	85.0	80.9	78.2	74.7	66.6
100	104.1	95.2	90.8	87.6	84.0	75.3

Appendix B
High-usage leased circuits with overflow to the public telephone network

See section 8.1 for discussion. This table is based on typical criteria which may not accord with some situations encountered in practice. The effect for small circuit numbers will not be significant but, where reductions in excess of five circuits are indicated, it may prove worthwhile to test the effect on costs of varying the suggested figure.

Calculated circuit requirement for wholly leased-circuit working	Equivalent circuit requirement for high-usage working
1	Not applicable
2	1
3	2
4	3
5	4
6	5
7	5
8	6
9	7
10	7
11	8
12	9
13	9
14	10
15	11
16	12
17	13
18	13
19	14
20	15
21	15
22	16
23	16
24	17
25	18

Glossary

Alternative routing (in US often **alternate routing**)   A method of routing traffic from a switching centre which offers an automatically controlled choice between two or more possible routings. The two principal variants are (a) fixed (see Section 7.5, Fig. 7.4a and Section 8.1), where the possible routings – usually two in number – are predetermined right through to the destination switching centre and a second-choice routing is selected only when all circuits on the first-choice routing are busy, and (b) dynamic (see Section 7.5, Fig. 7.4b and Section 8.1), where the first-available free circuit on any possible routing is selected stage by stage, by each switching centre in turn, until the required destination is reached.

American Standard Code for Information Interchange (ASCII) (pronounced ASKEY)   A standard US data-communications code for the *synchronous transmission* of data. See also *International Alphabet no. 5*.

Amplitude   The characteristic of a transmitted electrical waveform which equates to signal strength and hence speech volume. Amplitude decreases with the distance transmitted as a result of *transmission loss*.

Analogue transmission   The transmission of information – speech, pictures or data – by means of an continuous electrical waveform. The frequency variations, but not necessarily the actual frequencies transmitted, match those of the original speech, scanned-picture or data waveform (see also *audio transmission* and *frequency-division multiplexing*).

Analytical model   See *mathematical model*.

Asynchronous (data) transmission   Sometimes referred to as start-stop transmission, since each character transmitted is preceded by a start signal and followed by a stop signal; the receiver is thus started and stopped for each character. Used with *International Alphabet no. 2* for telex. Transmission spends up to $\simeq 1200$ bit/s possible over an analogue circuit or $\simeq 19.2$ kbit/s digital. See also *synchronous (data) transmission*.

Audio transmission   A specific case of *analogue transmission* where the electrical frequencies transmitted (normally between 300 and 3400 hertz) match those of the speech waveform (see *hertz*).

Automatic repeat attempt   See *repeat attempt*.

Availability Within a telephone switching centre, describes the extent to which a group of circuits – for example, a *traffic route* – can be accessed from a preceding circuit. This influences the number of circuits needed on the route. Modern electronic switching systems offer full availability, and the *traffic table* of Appendix A is calculated on this basis. The switch design of electromechanical systems permitted only limited availability, defined by the maximum number of accessible circuits – typically 10 or 20.

Baseband LAN See *local area network*.

Basic reference model See *open systems interconnection*.

baud A unit of data-modulation rate, equal to one signal element per second (not necessarily the same as a *data transfer rate* of one bit per second: see Section 3.2 and Fig. 3.2), which dates from the early days of telegraphy and is still applied to telex operation.

Binary notation The basis of codes used for data transfer, in which numbers are represented by combinations of two elements 0 and 1.

bit Abbreviation of 'binary digit': the smallest separable item of data that can be transmitted, equal to one electrical or optical pulse. *Data transfer rate* and digital *transmission speed* are measured in bits per second (bit/s).

Block See *data block*.

Blocking Is experienced when a free path cannot be found through a switching centre, resulting in a failed attempt to transfer information. Most modern switching systems are designed to be non-blocking.

Branching connection The connection between a *station* situated off the main transmission path of a data communications network, especially a *local area network*, and that path.

Bridge See *data-link bridge*.

Broadband circuit See *wideband circuit*.

Broadband LAN See *local area network*.

Broadcast operation See *wide area network*.

Buffer A temporary store for data prior to its transmission over a link, which permits the more efficient use of link circuits, changes in *transmission speed* and/or retransmission when errors are detected.

Burst-mode transmission A technique which enables a continuous flow of digital information to be exchanged over a single cable pair or equivalent radio channel by the transmission of short, high-speed bursts of data in each direction alternately, thus avoiding the problem of mutual interference (see Section 4.1 and Fig. 4.1a). Used specifically to provide digital *local-circuit* access to the *integrated-services digital network*. See also *echo-cancelling transmission*.

Bus A linear data-communications network *transmission path*.

Busy hour The hour of each working day during which the *traffic flow* over the whole or part of a telecommunications network can be expected to attain its highest level.

Busy-hour call attempt (BHCA) See *calling rate*.

byte A group of *bits* representing one character. Computers, originally designed to process 8-bit bytes as transmitted over networks (see *data octet*), may now use 16-bit or 32-bit bytes internally.

Call holding time   See *holding time*.

Calling rate   The rate of origination of calls over a telephone network, expressed as the average number of *busy-hour* call attempts per *local circuit* or PBX extension. A calling rate of 0.5 means that, on average, one call attempt is originated over each local/extension circuit every other busy hour. Calling rates are an essential input to switching-centre and PBX *dimensioning* and the maximum call handling rate of PBXs is quoted in terms of *busy-hour call attempts* (BHCA). Public-network calling rates are calculated separately for business and residential users for each *local switching centre*.

Cambridge ring local area network   See *empty-slot data-transfer technique*.

Carried traffic   The amount of traffic carried by, for example, a traffic route; used with the hypothetical quantities of *lost traffic* and offered traffic in telephone-network traffic calculations.

Carrier-sense multiple-access (CSMA) data-transfer technique   Originally developed by the US Xerox Corporation for the proprietary *Ethernet local area network*. In CSMA with collision detect (CSMA/CD), the added feature prevents the clashing of simultaneous transmissions by different stations. See Section 5.3 and Fig. 5.8

Carrier (transmission) system   Now largely outdated term for a *transmission system* that enabled a number of circuits to be provided over two cable pairs by means of *frequency-division multiplexing*. See *FDM (transmission) system* (Section 2.2).

Catchment area   Of a *local switching centre*, the area bounded by the furthermost users directly connected to it by *local circuits*. The catchment area of a *trunk/toll switching centre* embraces the catchment areas of the local switching centres served.

CCIR, CCITT   See *International Telecommunications Union*.

CCS   See *hundred call seconds per hour*.

Cellular (mobile) radio network   A mobile ratio network serving a geographical area that is divided into cells. Each cell has its own short-range transmitter/ receiver base station, operating on a different frequency from those in adjacent cells; this enables a limited range of frequencies to serve an unlimited area (see Section 2.9 and Fig. 2.11).

Centralised processor control   Of a *stored-program-control switching centre*, where control is centralised in a single *processor* or *multiprocessor* group.

Central office   US term for *local switching centre*.

Centrex   *Private branch exchange* (PBX) services offered from a public-network *local switching centre*, needing no on-site PBX equipment.

Channel   A *transmission path* over a cable or radio *transmission system*. Each channel provides the equivalent of two wires and is capable of transmission in one direction only; thus two channels are needed per *circuit*. See also *FDM (transmission) system (analogue transmission)* and *TDM (transmission) system (digital transmission)*.

Channel-associated signalling   See *signalling*.

Circuit   An assembly of electrical/electronic components and wiring designed for a particular purpose; for example, a functional element of a switching centre, as in *line (termination) circuit*, signalling circuit, etc. In telecommunications practice, the term is also applied to a communication path between two components of a network: hence *local circuit, traffic-route* circuit, point-to-point circuit, two-wire circuit, four-wire circuit, *duplex (data) circuit*, etc.

Circuit occupancy   See *occupancy*.

Circuit switching   A method of transferring *telecommunications traffic* between users by means of the temporary linking of independent point-to-point *circuits* between *switching centres*, where each call is treated as an independent transaction, as used for telephone networks. See Section 3.1.

Clear-down   The process of disconnecting the temporarily established communication path between two users on completion of a call over a circuit-switched network. See *circuit switching*.

Coaxial cable   A cable made up of one or more concentric pairs of conductors, each pair consisting of a central copper 'wire' surrounded by, but insulated from, an outer copper tube. This form of pair construction permits very high frequencies to be transmitted with much less *transmission loss* than over a conventional pair of wires.

Common-channel signalling   See *signalling*.

Computer-aided design (CAD)   The use of computer-based design aids to speed system design, ensuring a uniform, structured approach to electronic and software design problems.

Computer model   See *computer simulation*.

Computer simulation   Involves the writing of a computer program which is a computer-language precis of the performance-influencing events that occur during the operation of, for example, a telecommunications or computing system, as a means of evaluating the performance of that system; the program is termed a computer model. See also *performance evaluation*.

Concentration, concentrator   See *traffic concentration point*.

Congestion   The term used to describe the situation that obtains when *telecommunications traffic* cannot be transferred to its destination because of a shortage of circuits *en route*. The network or a particular switching centre or traffic route is then said to be congested.

Conversational traffic   During a successful call over a telephone network, describes the traffic during that part of the call when the users are in contact and information, for example speech, can be exchanged. The elements of call *holding time* represented by call *set-up* and *clear-down* are excluded. See Section 1.1

Cross-network delay   Data-transfer delay as measured end-to-end across a network. See also *message transit time* and *response time*.

Data block   A group of *data octets* or *bytes* transmitted consecutively and processed as a single entity.

Data frame   A *data packet* to which control and error-correction data has been added for transmission over a packet-switched network *link*. See Section 3.3.

Datagram working   See *packet switching*.

Data-link bridge   A device for interfacing a *local area network* to a *link* in a *wide area network*. See Section 5.3 for operation.

Data octet   A group of 8 *bits* representing one character, transmitted over a telecommunications network as a unit (see also *byte*).

Data packet   A fixed number of message-carrying *data blocks* assembled with *header* (sequence, identity and destination information) for transfer over a packet-switched network. See Section 3.3 and Fig. 3.4

Data throughput   See *throughput*.

Data transfer rate   The rate at which data is transferred from one point to another, measured in bits per second (bit/s).

Decentralised processor control   See *distributed processor control*.

Delay working   A method of circuit seizure used in circuit-switched networks; if all circuits to the required destination are engaged, *set-up* is automatically delayed in a queue until a circuit becomes free. See also *loss working*.

Digital access signalling system (DASS)   Interim UK standard system (pending agreement on an international standard) for signalling over a public-network digital *local circuit* to the *integrated-services digital network*.

Digital private-network signalling system (DPNSS)   As for *digital access signalling system*, but for signalling over private networks.

Digital transmission   The transmission of information – speech, pictures or data – in the form of coded data *bits* (see also *pulse-code modulation, time-division multiplexing* and Section 2.4).

Dimensioning   In telecommunications practice, the process of calculating how many switching-system modules or traffic-route circuits are needed to carry or process a particular forecast traffic load.

Discounted cash flow (DCF)   A process of comparing the costs and benefits which are expected to accrue from alternative courses of action to achieve a particular objective, which takes into account the value of money relative to time. See Section 11.3 and example in Section 11.4.

Distributed processor control   Of a *stored-program-control switching centre*, where control is distributed among a number of *processors* (usually *microprocessors*), each located with the equipment it controls.

Duplex or full-duplex (data) circuit   A circuit capable of transmitting data in both directions simultaneously. See also *simplex* and *half-duplex*.

Dynamic alternative routing   See *alternative routing*.

Echo-cancelling transmission   A technique which enables a continuous flow of digital information to be exchanged over a single cable pair or equivalent radio channel in both directions simultaneously, by the generation of signals which cancel out echoes that would otherwise cause mutual interference (see Section 4.1, Fig. 4.1). Used specifically to provide digital *local-circuit* access to the *integrated-services digital network*. See also *burst-mode transmission*.

Empty-slot data-transfer technique   Originally developed by Cambridge University for the *Cambridge ring local area network* (LAN); now used extensively for various proprietary *slotted-ring local area networks* (see Section 5.3, Fig. 5.9).

End-to-end grade of service   See *overall grade of service.*

erlang   The internationally-agreed unit of telecommunications traffic flow: unit symbol E. A circuit in continuous use for 1 hour is said to be carrying 1 E of traffic flow; if it is in use for only 50% of the hour, the traffic flow is 0.5 E. See Sections 1.3 and 6.2 and Fig. 6.1.

Ethernet local area network   See *carrier-sense multiple-access data-transfer technique.*

Event   In computing and telecommunications practice, any point in a computer program where an external action is caused to occur or where an external action initiates a particular program sequence. Events occur at specific time intervals relative to each other.

FDM (transmission) system   A transmission system that enables large numbers of circuits to be provided over two *coaxial-cable* pairs, one for each direction of transmission, by means of *frequency-division multiplexing.* Systems are described in terms of the highest frequency transmitted (up to 60 MHz) and system capacity is stated in terms of the number of *channels* provided for one direction of transmission (which equates to the number of circuits). See Section 2.2.

Fixed alternative routing   See *alternative routing.*

Flow control   The control of the transmission of *data blocks* to a store-and-forward receiving terminal to ensure that its storage capacity is not exceeded.

Frame   See *data frame.*

Frequency-division multiplexing (FDM)   Several *audio transmission* channels are each modulated to a different, progressively higher frequency range so that all may be transmitted over a single *transmission path.* See *modulation* and Section 2.2.

Full availability   See *availability.*

Fully connected   See *mesh network.*

Fully-provided (traffic) route   A traffic route, the number of circuits of which has been calculated to a *grade of service* of 0, that is, sufficient circuits have been provided to ensure that there will be no call failures due to a shortage of circuits at the forecast *busy-hour* traffic level. See also *high-usage (traffic) route.*

Grade of service (GOS)   A measure of the quality of service provided at each stage throughout a *circuit switched* telephone or telex network, in terms of the sufficiency of circuits or switching-centre equipment to carry a given level of telecommunications traffic. GOS is specified in terms of the probability that a given proportion of call attempts will be lost during the *busy hour.* It is usually expressed as a decimal; for example, a GOS of 0.02, when applied to a *traffic route*, means that 1 call attempt in 50 can be expected to fail to find a free circuit on that route during the busy hour. The GOS concept is also applicable to *message switched* telex networks, where it relates to message attempts instead of call attempts.

Half-duplex (data) circuit   A circuit capable of transmitting data in either direction, one direction at a time. See also *duplex.*

Handover (US handoff)   See *in-call handover*.

Header   Of *packet switching*, the sequence, identity and destination data added to a group of message-carrying *data blocks* to form a *data packet*.

hertz The internationally agreed unit of electrical frequency, equal to 1 cycle per second: unit symbol Hz. Frequencies often need to be expressed in kilohertz (1 kHz = 1000 Hz), megahertz (1 MHz = 1000 kHz) or gigahertz (1 GHz = 1000 MHz).

Hierarchical network   A telecommunications network in which the switching centres are organised in hierarchical tiers; traffic for local destinations remains in the lowest tier and that for more distant destinations is progressed upwards through the higher tiers as necessary.

High-usage (traffic) route   A traffic route, the number of circuits of which has been calculated to a very poor *grade of service* (GOS) to ensure the maximum possible circuit usage. With fixed alternative routing, a first-choice route is commonly high usage; any excess traffic is diverted to a second-choice, *fully provided route* (in a private network the second-choice routing may be the public switched network). See *alternative routing*.

Holding time   The time that a circuit or item of telecommunications equipment is engaged for the purpose of transferring, or attempting to transfer, information between users. The holding time for a call over a telephone network, from the commencement of *set-up* to *clear-down*, is specifically referred to as call holding time.

hundred call seconds per hour   A US alternative unit to the *erlang* for the measurement of telecommunications traffic flow: unit symbol CCS. The time a circuit is in use is measured in multiples of 100 seconds instead of as the decimal part of an hour; hence a circuit in continuous use for 1 hour is said to be carrying 36 CCS = 1 erlang.

In-call handover (US handoff)   In a *cellular (mobile) radio network*, the process which takes place as a mobile user traverses a boundary between cells whilst a call is in progress. The *transmission path* is automatically transferred from the base station of the cell being vacated to the base station of the cell being entered, without interrupting the call. See Section 2.9 and Fig. 2.12

Integrated digital network (IDN)   A telephone network in which all switching centres and interconnecting transmission links use *digital transmission* techniques, *analogue transmission* being retained only on *local circuits*. Speech and other information is encoded digitally on entry to the originating *local switching centre* and remains in this form until it leaves the destination local switching centre. See Section 2.8.

Integrated-services digital network (ISDN)   An *integrated digital network* which has been enhanced by the provision of *digital transmission* over *local circuits*. Digital transmission end-to-end between users enables the telephone network to carry an extended range of data and other services. See Section 4.1 and Fig. 4.2.

Integrated-services PBX (ISPBX)   A PBX that provides speech and data integration over common extension circuits together with public *integrated-services digital network* compatiblity. See Section 5.4.

International Alphabet no. 2 (IA2) A standard CCITT data-communications code for the *asynchronous transmission* of data, especially telex. See Section 3.2.

International Alphabet no. 5 (IA5) A standard CCITT data-communications code for the *synchronous transmission* of data, virtually identical to the *American Standard Code for Information Interchange*. See Section 3.1.1.

International Telecommunications Union (ITU) A specialist agency of the United Nations, based in Geneva and responsible for international telecommunications standards. Detailed work is divided between two main committees, the International Telegraph and Telephone Consultative Committee (CCITT) and the International Radio Consultative Committee (CCIR). See Section 2.3.

I series recommendations (of CCITT) The CCITT recommended standards relating to the *integrated-services digital network* (ISDN). Most of these are concerned with establishing internal standards for the ISDN, but the following, which relate to the *network terminating equipment* (see also Section 4.1), are of particular relevance to ISDN users:

I420 user-to-network interface requirements for single-line access
I421 user-to-network interface requirements for multiple-line access (1.5 or 2.0 Mbit/s)
I461 X21/X21 *bis* data terminal access
I462 X25 (*packet-mode*) data terminal access
I463 *V series* data terminal access.

Key system A small *private branch exchange* equipped with special telephones, from which each extension obtains access to the public network or other extensions by the operation of a separate keyswitch for each line rather than by dialling or keying a number.

Leased private circuit A circuit leased from a public telecommunications network operator to provide a communications path between two fixed locations for the private use of the lessee. Circuits may be leased on a permanent, short-term or recurring-time basis.

Limited availability See *availability*.

Line concentrator A device which enables several public-network telecommunications users to be connected to a *local switching centre* over a lesser number of *local circuits*, which are shared on a time basis.

Line (termination) circuit An element of a *local switching centre*, required on a per *local circuit* basis to permit the exchange of signals between a user on that circuit and the switching centre.

Link A communications path linking two components of a network, particularly a data-communications network.

Link transmission or propagation time Electrical or optical signals are propagated along a link at the speed of light; this takes a very short, measurable time which may need to be taken into account in circumstances where a rapid data-network response is vital. See *message transit time* and *response time*.

Local area network (LAN) A self-contained private data-network system which provides for the transmission and switching of packetised data between data terminals, computers and printers, etc. in a confined geographical area. There are two basic versions, the baseband LAN with a *bus* or ring network structure, and the broadband LAN. The latter uses the *tree network* structure and components of cable television practice, and has the capability of providing a parallel telephone service as well as several independent data services. See *packet switching*; see also Section 5.3 and Figs. 5.6 and 5.7 (baseband) and 5.11 (broadband).

Local circuit   The normally two-wire circuit between a public telecommunications service user's premises and the *local switching centre*.

Local network   The network of cables and radio links over which *local circuits* are provided.

Local switching centre   A switching centre to which public telecommunications users, particularly of a telephone network, are directly connected via *local circuits*. See also *trunk/toll switching centre*.

Loss working   A method of circuit seizure used in circuit-switched networks. If all circuits to the required destination are engaged, the call attempt fails and is said to be lost, and a fresh call attempt is then required if the desired connection is to be established. See also *delay working*.

Lost traffic   A hypothetical quantity used in telephone-network traffic calculations, derived from the number of failed call attempts due to a shortage of circuits multiplied by the average call *holding time*. Lost traffic plus *carried traffic* equals the similarly hypothetical offered traffic. See Section 1.2.

Mathematical model   One or more equations which are a mathematical precis of the performance-influencing events that occur during the operation of, for example, a telecommunications or computing system. The model is used to evaluate the performance of that system. See also *performance evaluation*.

Mesh network   A telecommunications network in which locations that require to communicate with each other (for example, the *stations* of an *unswitched data network*, or *switching centres*) are directly linked. Where every location is linked to every other location, the network is said to be fully connected; where only some of the possible links are provided (this requires some pairs of locations to communicate via another intermediate location, hence *tandem switching* is essential), the network is said to be partially connected.

Message handling time   The time taken to switch a data message through a message-switched network *switching centre*. See *message switching* and *message transit time*.

Message pair   A message from a data terminal to the central computer of an *unswitched data network* plus the response message returned by the computer to the terminal. By determining the average overall time taken for such a transaction, it is possible to estimate *traffic flow* over a network link in terms of the number of message pairs of average duration in a given period.

Message switching   A method of transferring data traffic between users by means of the temporary linking of independent point-to-point *circuits* between *switching centres*, where each undirectional message is treated as an independent transaction, as used for telex networks. See *store-and-forward working* and Section 3.2.

Message transit time   The time taken to transmit a message across a message-switched data network from sender to receiver; the sum of switching-centre *message handling time(s)* and *link transmission times*.

Microprocessor   A processor, of limited processing power compared with a large mainframe computer, which largely consists of a purpose-designed silicon chip and is therefore physically very small. The microprocessor is widely used for the control of telecommunications switching functions: see Section 7.4.

Microwave radio (relay) system   A point-to-point radio *transmission system* capable of carrying large numbers of telecommunications circuits, using *analogue transmission* or *digital transmission* techniques. See Section 2.6.

Mobile radio network   A telecommunications network which uses radio communication between mobile users and fixed base stations. The base stations of a public mobile radio network are linked to the public telephone network to permit communication between mobile and fixed-location users. See *cellular (mobile) radio network* and Section 2.9.

Modem   A combined modulator and demodulator, necessary between a data terminal and an analogue *leased private circuit* or telephone-network *local circuit* to reduce the probability of errors at transmission speeds greater than a few hundred bits per second. A modem is designed to operate at a specific transmission speed, transmitting digital bits to line as audible tones and presenting received tones to the terminal as bits. See also *modulation* ((b) frequency and (c) phase) and Section 3.1.1.

Modulation   The process of altering the characteristics of a varying electrical signal to facilitate its transmission over a telecommunications circuit. Four principal techniques are relevant to transmission methods described in this book; of these, (a), (b) and (c) are used for *analogue transmission* whereas (d) converts an analogue signal for *digital transmission*:

(a) Amplitude modulation, where the signal waveform is electrically lifted by being imposed directly on to a higher carrier frequency, as in *frequency-division multiplexing*.
(b) Frequency modulation, where the signal waveform is substituted by different frequencies; for example, the elements of a data signal in a *modem* (hence the term 'frequency-shift keying' (FSK)).
(c) Phase modulation, a process also used in a data *modem* to achieve essentially the same ends as frequency modulation (hence the term 'phase-shift keying' (PSK)).
(d) *Pulse-code modulation*, a development of pulse-amplitude modulation which is itself based on the principle of (a) above, where the signal waveform is substituted by coded pulses.

Multidrop data network   An *unswitched data network*, the shared transmission path of which serves a number of *stations* connected at intervals along its length. See Section 5.3 and Fig. 5.5b.

Multi-frequency (MF) signalling   A method of signalling a telephone number in which each digit is represented by two different simultaneously transmitted tones. Commonly used to provide faster signalling between a keyphone and an electronic *private branch exchange* or *local switching centre.*

Multiple addressing   In data communications, describes the process necessary to send a message to a number of destinations simultaneously.

Multiplexing   A means of providing a number of independent transmission paths over a single cable pair or equivalent radio link. See *frequency-division multiplexing, time-division multiplexing* and, for packet-switched data, *statistical time-division multiplexing.*

Multipoint radio system   A microwave radio system designed to provide digital *local-circuit* facilities for a number of telecommunications users located within a limited radius and distance, where *integrated-services digital network* access is required and it is uneconomic or impossible to provide this over metallic pairs.

Multipoint data network   See *multidrop data network.*

Multiprocessor   A group of several independent processors linked together to provide the processing power needed for the *centralised processor control* of a large public-network switching centre.

Net data throughput   See *throughput efficiency.*

Network delay   See *cross-network delay.*

Network operator   Used in this book to mean the administration or company that owns and operates a telecommunications network.

Network terminating equipment (NTE)   The equipment required at the user end of a digital *local circuit* to provide the additional functions necessary for *integrated-services digital network* access. See *burst-mode transmission, echo-cancelling transmission* and Section 4.1.

Non-blocking   See *blocking.*

Occupancy or **circuit occupancy**   A measure of the time a telecommunications *circuit* is engaged or occupied during the *busy hour*, expressed as a decimal figure with a maximum of 1.0. See Section 1.3.

Octet   See *data octet.*

Offered traffic   See *lost traffic.*

Open systems interconnection (OSI)   The internationally sponsored concept designed to achieve a single integrated set of standards for data communication through the application of the OSI *basic reference model* (IS7498). See Section 3.4 and Fig. 3.7.

Optical transmission   The transmission of information over an optical fibre in the form of pulses of light, using *digital transmission* techniques. See Section 2.5.

Overall grade of service (GOS)   The overall effect on quality of service of the individual *grade of service* standards applied to the switching centres and interconnecting traffic routes of a telephone or telex network, as perceived by a user; sometimes referred to as the end-to-end GOS. See Sections 6.5.1 and 10.2.

Packet See *data packet.*

Packet assembler/disassembler (PAD) A device that may be interposed between a user's standard data terminal and line, to packetise data for transmission over a packet-switched network and translate received packets into a continuous data stream. See *data packet, packet mode terminal* and *packet switching.*

Packet handling time The time taken to switch a data packet through a packet-switched network switching centre. See *packet switching* and *response time.*

Packet mode terminal A specially designed data terminal that is able to packetise data for transmission over a packet-switched network and translate received packets into a continuous data stream, without the need for a separate *packet assembler/disassembler.* See *data packet* and *packet switching.*

Packet switching A method of transferring data traffic between users where each unidirectional message is cut at fixed intervals into extremely short *data packets*, each of which is treated as an independent transaction. The *store-and-forward* principle is used to transfer packets over a time-shared *transmission path*, either via linked public-network-type digital *switching centres* or over a *local area network.* The former employ dynamic *alternative routing* and either datagram working, in which each packet is transmitted and switched independently as a separate self-routing transaction, or virtual-call working, in which the same routing is used for all packets constituting a single message. See Sections 1.9 and 3.3.

Parent Of a *local switching centre*, that to which the particular user is directly connected via a *local circuit.*

Partially connected See *mesh network.*

PCM (transmission) system A system that utilises *pulse-code modulation* and *time-division multiplexing* to provide 24 (US) or 30 (European) *circuits* – each comprising two unidirectional 64 kbit/s *channels* – over two cable pairs.

Performance engineering In telecommunications and computing practice, the process of carrying the results of *performance evaluation* into system and network design, to achieve the best possible *traffic-carrying/handling performance* at minimum cost. See Section 9.1.

Performance evaluation In telecommunications and computing practice, the process of predicting the *traffic-carrying/handling performance* capability of system and network designs and, where feasible, measuring the performance of live systems and networks. See Section 9.2.

Point-to-point operation See *wide area network.*

Point-to-point radio system A microwave radio system designed to provide digital *local-circuit* facilities for a number of telecommunications users, where *integrated-services digital network* access is required and it is uneconomic or impossible to provide a link between the two points over metallic pairs. The term also applies generally to any radio system designed for point-to-point operation.

Polling mechanism   The general term used to describe techniques for regulating access to a computer over the shared transmission path of an *unswitched data network*, from a number of data terminals.

Private branch exchange (PBX)   A *switching centre* which enables the telecommunications users of an individual business enterprise, usually located on a single site, to communicate with each other and with the public telephone network. See Section 5.2.

Private telecommunications network   A network which, although it may provide access to and from public telecommunications networks, is paid for by, and intended to serve the communications needs of, an individual business enterprise.

Processor   The term appied to a computer used to control a telecommunications *switching centre*. See *centralised processor control, distributed processor control, stored-program control* and Section 7.4.

Processor load control   The process of regulating the traffic load on a telecommunications *switching centre* to ensure that its controlling *processor(s)* cannot fail as a result of being overloaded. See Section 7.4.

Protocol   In data communications, a set of rules which defines the conditions necessary for data to be exchanged between two points.

Protocol conversion   In data communications, the process of translation from one *protocol* to another, necessary to enable two users to exchange data between equipments which normally operate to different protocols.

Public switched telephone network (PSTN)   See *public telecommunications network.*

Public telecommunications network   A publicly or privately owned and operated network which provides the facility for customers to communicate with each other and with the customers of other public telecommunications networks on a fee-paying basis.

Pulse-code modulation (PCM)   The process by which an analogue electrical waveform is substituted by coded pulses to enable *digital transmission*. See Section 2.4 and Fig. 2.3.

Registration   In a *cellular (mobile) radio system*, the process which records the cell in which each mobile radio user is currently located, so that calls to mobile users can be directed to the appropriate base-station transmitter.

Repeat attempt   An attempt to complete a desired telephone or telex network connection which closely follows a failed attempt to complete the same connection. See Section 1.4.

Repeater   In a telecommunications network, any device which repeats a received signal for onward transmission.

Response time   In data communications, the time between sending a signal from a terminal to a computer and receiving a response from that computer. Where terminal and computer are linked via a network, response time provides a measure of network as well as computer performance, the network component being made up of *link transmission time(s)* and, where appropriate, *packet handling time(s)*.

Route See *traffic route*.

Router A device for interfacing a *local area network* to a *link* in a *wide area network*. See Section 5.3 for operation.

Routing See *traffic routing*.

Service-independent network The as yet theoretical concept of a switched telecommunications network capable of transporting any conceivable telecommunications service. See *variable-bit-rate switching*; see also Section 4.2 and Fig. 4.4.

Set-up The process of establishing a communication path between two users over a circuit-switched network. See *circuit switching*.

Set-up delay The time taken to set up a connection between two users of a circuit-switched network. See *set-up*.

Signalling The process of transmitting/receiving coded information, particularly applied to (a) data communications, and (b) control and routing information sent over a telephone network to set up, meter for charging purposes, change the status of and clear down a call. Inter-switching-centre signalling information may be transmitted over the selected communication path, termed channel-associated signalling, or over a separate signalling path in company with the signalling information for other calls, termed common-channel signalling. See Section 2.8, Fig 2.7 and Fig. 2.10.

Simplex (data) circuit A circuit capable of transmitting data in one direction only. See also *duplex* and *half-duplex*.

Simulation model See *computer simulation*.

Single-choice routing The simplest method of routing traffic from a switching centre, providing only one possible routing to any particular destination; if, at any stage along this routing, all circuits on the *traffic route* concerned are engaged, the attempt to establish a connection fails. See Section 7.5, Fig. 7.3 and Section 8.1.

Sizing In computing practice, the process of calculating link transmission speeds and the number of computer/ancillary-equipment modules needed for a particular forecast computing load.

Slotted-ring local area network (LAN) A ring-based *local area network* of the baseband type which uses the *empty-slot data-transfer technique*. See Section 5.3 and Fig. 5.9.

Space-division switching A telecommunications switching method in which a metallic path is switched between a number of spatially divided destination points. See Section 2.8; see also *time-division switching*.

Speech-transmission frequency range The frequency range it is possible to transmit over one *channel* of a *transmission system* where *analogue transmission* is employed: normally 300–3400 Hz. A 64 kbit/s *digital transmission* channel provides an equivalent capability.

Star-connected ring A network configuration which may be used to provide added security against link failure for a ring-based *local area network*. See Section 5.3 and Fig. 5.6b.

Star network   A network in which telecommunications or computer users are each connected to a centrally located hub where, for example, the hub may be a *local switching centre* to enable telephone users to communicate with each other, or a computer with which data-terminal users wish to communicate (see *unswitched data network*).

Start-stop transmission   See *asynchronous (data) transmission.*

Station   (a) The site of a telecommunications transmission facility, as in repeater station, radio station, etc. (b) In data communications, one or more data terminals etc, which are co-located and share a common transmission path giving access to a data network.

Statistical (time-division) multiplexing   The process of interposing *data packets* from various sources prior to transmission over a link. See *packet switching* and Section 3.3.

Store-and-forward working   Derives economic advantage from the fact that data may be delayed in transit without affecting the intelligibility of the information being transferred. Storage capacity is provided to allow (a) data awaiting transmission to be automatically held awaiting a free transmission path and (b) excess received data to be held awaiting processing; traffic peaks are thus absorbed in the storage and changes in transmission speed can be accommodated. Store-and-forward working is inherent to the *packet-switching* concept and may also be used in conjunction with *message switching*: see Section 1.8 and Fig. 1.6. Data terminals are also obtainable with a store-and-forward capability.

Stored-program control (SPC)   Of a *switching system*, where the control functions are the responsibility of a computer program.

Switching centre   An interconnecting point in a telecommunications network, built up from *switching-system* equipment modules, which enables connections between the network *links* to be established and disconnected automatically as required by the users of that network. See also *local switching centre, private branch exchange* (PBX) and *trunk/toll switching centre.*

Switching system   An integrated range of equipment modules designed for assembly as a *switching centre*. See Section 7.1.

Synchronous (data) transmission   A method of transmitting data in which synchronising *bits* are interposed at intervals between the message-carrying bits to ensure that the receiver remains in step with the transmitter. Used typically with the *International Alphabet no. 5*; this permits higher transmission speeds than *asynchronous (data) transmission.*

Tandem switching   Describes the situation where two switching centres or PBXs are linked via a third, intermediate switching centre.

TDM (transmission) system   A transmission system that enables large numbers of circuits to be provided over two *coaxial-cable* pairs, one for each direction of transmission, by means of *time-division multiplexing*. Systems are described in terms of the pulse transmission speed or *bit* rate, usually in Mbit/s, and system capacity is stated in terms of the number of *channels* provided in one direction of transmission (which equates to the number of circuits). TDM system capacity is built up from multiples of 24 channels (US) or 30 channels (European). See *PCM transmission system*; see also Section 2.4.

Telecommunications traffic Information – speech, pictures or data – conveyed by electrical or optical means between telecommunications users. For the purpose of traffic calculation, a circuit in a *circuit-switched* (telephone) network is said to be carrying traffic when it is engaged and unavailable for any other call; that is, from the commencement of *set-up* to the end of *clear-down*. See also *conversational traffic* and Section 1.1.

Throughput The amount of data that can be processed in a given amount of time, measured in bits, packets, etc. per second as appropriate.

Throughput efficiency or **net data throughput** The efficiency of a data network in terms of the percentage of usable information bits received relative to the total amount of data transmitted. See Section 8.2.

Time-division multiplexing (TDM) The coded electrical or optical pulses which represent information to be transmitted digitally are speeded up prior to transmission so that each pulse is reduced in duration; groups of pulses for different *channels* can then be interposed and transmitted over a single *transmission path*. See *digital transmission, pulse-code modulation*; see also Section 2.4 and Fig. 2.3.

Time-division switching A method of switching digital telecommunications traffic. The successive groups of pulses for both channels of a *digital transmission* circuit are transferred from their allotted time slot of the incoming transmission system to a different time slot allocated to the required circuit on an outgoing transmission system. The pulses are effectively switched in time by being momentarily delayed, instead of in space as with *space-division switching*. See *time-division multiplexing*; see also Section 2.8 and Fig. 2.8.

Token-access/token-passing data-transfer technique Originally developed by the US International Business Machines Corporation (IBM); now used extensively for their own and other proprietary *token-bus/token-ring local area networks*. See Section 5.3 and Fig. 5.10.

Token-bus/token-ring local area network (LAN) Bus- or ring-based *local area networks* of the baseband type which use the *token-access/token-passing data-transfer technique*. See Section 5.3 and Fig. 5.10.

Toll switching centre US equivalent of *trunk switching centre*.

Traffic See *telecommunications traffic*.

Traffic-carrying performance The efficiency with which a telecommunications circuit, system or network is able to carry its designed *traffic flow*. See also *traffic-handling performance*.

Traffic concentration point/concentrator A point in a telecommunications system or network at which traffic from a number of sources is concentrated on to a smaller number of outlets so as to use the latter more efficiently.

Traffic engineering The process of designing and/or dimensioning a telecommunications system or network to achieve the optimum *traffic-carrying/traffic-handling performance*.

Traffic flow   The succession of telephone calls, data messages or other information passing over a telecommunications circuit, measured in terms of the proportion of time the circuit is in use. See *erlang, hundred call seconds per hour* (CCS) and *occupancy*.

Traffic-handling performance   A term often used synonymously with *traffic-carrying performance*, but which strictly ought to be applied only where traffic is processed, as by a *switching centre*, rather than simply carried, as by a *circuit*.

Traffic route   One or more circuits between two *switching centres*.

Traffic routing   In a switched telecommunications network, the *traffic route* or succession of traffic routes which, when interconnected by one or more *switching centres*, enables a connection to be established from a given switching centre to a particular destination switching centre. See *single-choice routing, alternative routing*

Traffic table   A table which relates traffic-flow units (see *erlang* and *hundred call seconds per hour* (CCS)) to *circuits* for a given *grade of service*. See Section 6.5.1 and example at Appendix A.

Transceiver   A contraction of transmitter/receiver, used in *local area network* and radio terminology.

Transmission loss   The reduction in *amplitude* (loss in signal strength) and distortion that occurs with distance as an electrical or optical signal travels along a cable or through the ether.

Transmission network   The network of cables, *transmission systems* and *microwave radio systems* over which telecommunications *traffic routes* are provided (see Section 2.7).

Transmission path   Any path over which telecommunications signals may be transmitted, for example, a cable pair or a *channel* of a *transmission system*.

Transmission speed   In *digital transmission* and data communications generally, the speed or *bit* rate at which pulses are transmitted over a circuit, measured in bits per second (bit/s).

Transmission system   A transmission package designed to carry a fixed maximum number of *circuits* of one or more *traffic routes* over two cable pairs or radio transmission paths. See *carrier system, FDM system, PCM system, TDM system* and *microwave radio (relay) system*.

Transponder   A special *transceiver* used in cable television practice which has a particular application in broadband *local area networks*. See *tree network*; see also Section 5.3 and Fig. 5.11.

Tree network   A network consisting of a central linear transmission path with branching connections, especially as used for a broadband *local area network*, where all data is transmitted via a *transponder* at the base of the 'trunk' (the head end).

Trunk/toll switching centre   A *switching centre* which is designed to switch trunk (US toll) traffic between *local switching centres* in a public telephone network.

Unswitched data network   A network in which a large central computer is connected to a number of stations via a shared transmission path, with no means of switching between them. Access to the computer is regulated by a *polling mechanism*. See Section 5.3 and Fig. 5.5.

Variable-bit-rate switching A potential new variant of *time-division switching* which would allow the *bit* rate of the switched transmission path (currently 64 kbit/s) to be dynamically varied to suit each particular user's needs. See Section 4.2.

Virtual-call working See *packet switching*.

Voice-frequency (VF) signalling A method of signalling over the *channels* of an *analogue transmission* circuit, using voice-frequency tones. See Section 2.2.

V series recommendations (of CCITT) The CCITT recommended standards relating to the transmission of data over *analogue transmission* circuits. The following, which concern the interworking requirements between the network interface (in this case a *modem*) and a data terminal, are of particular relevance:

V21	300 bit/s asynchronous modem for connection to the public switched telephone network (PSTN)
V22/V22*bis*	600/1200 bit/s synchronous modem for connection to the PSTN or a leased public-network circuit
V23	600/1200 bit/s asynchronous modem for connection to the PSTN or a leased public-network circuit
V24	general, based on and virtually identical to RS232 (US Electronic Industries Association standard)
V25	PSTN automatic calling and answering
V26	1.2/2.4 kbit/s synchronous modem for connection to the PSTN or a leased public-network circuit
V27	2.4/4.8 kbit/s synchronous modem as for V26
V32	4.8/9.6 kbit/s synchronous mode as for V26

Wide area network (WAN) A group of widely dispersed *local area networks* which are interlinked to form a single integrated data network. The links are operated using either broadcast or point-to-point working; with the latter, interfacing between a LAN and its link to other LANs is either via data-link bridges or routers. See Section 5.3.

Wideband circuit A point-to-point telecommunications circuit which provides for the transmission of a wider analogue frequency range or faster digital bit rate than that available over a standard switched circuit, as required for the transmission of video signals.

X series recommendations (of CCITT) The CCITT recommended standards relating to the transmission of data over *digital transmission* circuits. The following, which (apart from X3) concern the interworking requirements between the network interface and a data terminal, are of particular relevance:

X3	*packet assembler/disassembler* (PAD)
X20	300 bit/s asynchronous (circuit-switched network)

X20*bis*   300 bit/s asynchronous for existing V24-based terminals
X21   up to 48 kbit/s synchronous (circuit-switched network)
X21*bis*   up to 48 kbit/s synchronous for existing V24-based terminals
X24   general, for circuit-switched network terminals
X25   general, for packet-switched network (*packet mode*) terminals
X28   PAD to asynchronous terminal
X29   PAD to packet mode terminal
X32   general version of X25 for dial-up access via telephone network

In addition, the X200 series of recommendations is concerned with the standards for *open systems interconnection*.

Index

Entries in **bold type** are also included in the **Glossary**

alternative or alternate routing, 46, 86, 111–113, 120–124
amplification, 15–17
amplitude, 15
analogue transmission, 14, 35, 58, 65
analytical model, 137
answer-back code (telex), 39
artificial traffic generator, 136
ASCII (American Standard Code for Information Interchange), 37
asynchronous transmission, 36, 41
auto-answer, 37–38
auto-dial, 37–38, 41
automatic repeat attempt, 6, 42
availability, 90
average call holding time or duration, 4, 79–80, 82–83, 123
average message holding time or duration, 84–85
average message-transit time, 128–146
average response time, 126, 129–130, 145–147
average traffic flow, 4, 79–80

baseband local area network (LAN), 67–72, 129–131, 147
base station (mobile radio), 31–33
Basic Reference Model (OSI), 48–49
Baud, 40–41
bearer circuit, 24
bit (data), 21, 36
bit/s, 21, 36, 74
block (data), 37, 44, 126–127
blocking, 4
bothway circuits, 3
branching connection, 67–69, 73, 130

bridge (data link), 74, 131
British Telecom, 54, 119, 149, 151–153
broadband local area network (LAN), 72–73, 131, 147
broadcast operation (wide area network), 74, 131
buffer, 43, 126, 128
burst mode transmission, 50–54
bus-type local area network (LAN), 69
busy or engaged circuit, 1
busy hour, 7–8, 80, 82–83, 89–91, 121–124
busy-hour call attempt (BHCA) rate, 80–81, 100, 106, 123
byte (data), 37

call attempts, 3–4, 80–81, 83, 100, 102, 106, 122–123
call holding time or duration, 2–5, 79–80, 82–83, 123
calling rate, 80–81, 100, 106, 123
Cambridge Ring (LAN), 70
carried traffic, 4
carrier-sense multiple access (CSMA), 69–70
  CSMA with collision detect (CSMA/CD), 69–70
catchment area, 115–117, 120, 149
CCIR (International Radio Consultative Committee), 18
CCIT (International Telegraph and Telephone Consultative Committee), 18, 47
CCS (hundred call seconds per hour), 5, 79–80, 82–85, 90
cellular mobile radio network/system, 31–33, 38
centralized (processor) control, 105–106, 108

centrex, 65
channel (transmission), 17
channel-associated signalling, 26
charging-group area, 149–150
circuit occupancy, 5, 84–85, 93–94, 122–123, 128–129, 146
circuit switching, 35, 40, 55, 57, 63, 75–78
clear-down, 2–4
coaxial cable, 17, 72–73
common-channel signalling, 27–30, 65, 76
computer-aided design (CAD), 135, 138
computer model, 133, 137–138
congestion, 4, 40, 78, 80, 84, 92, 119, 124
conversational traffic, 2
cost, 8, 87, 92, 96–97, 99–102, 104–105, 107, 111–112, 115–116, 118–124, 126–131, 132–135, 137, 139–153, 154, 156–176
cross-network delay, 11–13, 43, 78, 84–86, 92–94, 125–129, 131
cross subsidisation, 151
cycles per second, 14

data communication standards, 46–49
datagram working (packet switching), 43, 113, 129
data link bridge, 74, 131
data signalling code, 10, 37, 41, 49
data switching system, 103–105, 108–109, 113
data terminal, 9, 46–47, 125–126, 129
data traffic, 9–13, 24, 35–38, 52, 65, 84–87
data transfer rate, 40–41
decentralized (processor) control, 105–106
delay working, 5–6
design date, 142
digital access signalling system (DASS), 76
digital private network signalling system (DPNSS), 65
digital transmission, 19–23, 26–29, 35, 38–39, 45, 50–56, 58–60, 65, 76, 119
dimensioning, 93, 96, 123, 132–135
direct current signalling, 15, 17–18
direct dialling in (DDI), 150
discounted cash flow (DCF), 102, 160–162, 168–170
distortion, 15
distributed (processor) control, 105–106
distribution (mathematical), 89
duplex (data) circuit, 10, 127
dynamic alternative routing, 111–113, 123–124, 127, 129

echo-cancelling transmission, 51, 54
empty-slot data transfer technique, 70–71, 130
end-to-end grade of service (GOS), 91–92, 141–144
end-to-end signalling, 26
engaged or busy circuit, 1
equipment reliability, 92, 97, 105, 156
Erlang, 5, 79–80, 82–85, 90
error correction (data), 37, 42, 44–46, 49, 126–127
error rate (data), 37–38, 127
Ethernet, 70
event (computer program), 135
extension circuit/interface-unit (PBX), 62, 98–100

facilities (PBX), 61, 63
facsimile, 24, 38, 55, 152
fixed alternative routing, 111–113, 121–123
flow control, 126
forecast date, 142
forecasting (of traffic), 79–88, 118–119, 133, 141
four-wire circuit, 16–17, 26–29
four-wire switching, 26–28
frame (data), 37, 44–46, 85
frequency-division multiplexing (FDM), 17
    FDM transmission system, 17
front-end processor, 67
full availability, 90
fully-connected network, 109, 118
fully-provided (traffic) route, 90, 121

grade of service, 8–9, 84, 89–92, 101, 111–112, 121–122, 124, 128, 141–144

half-duplex (data) circuit, 10, 127
handover or handoff (cellular radio), 32–33
header (data packet), 44
Hertz, 14
hierarchical network, 118, 127, 128, 142
high-usage (traffic) route, 121–123, 144, 146
holding time: general, 2
    average, 4, 79–80, 82–85, 123
    call, 2–5, 79–80, 82–83, 123
    message, 11, 84–85
    packet, 13, 85–86
hundred call seconds per hour (CCS), 5, 79–80, 82–85, 90

in-call handover or handoff (cellular radio), 32–33

Institute of Electrical and Electronic Engineers (IEEE) (USA), 47
integrated digital access, 52
Integrated Digital Network (IDN), 27–31, 50, 55–56
Integrated Services Digital Network (ISDN), 52–56, 114
integrated services PBX, 76, 114
integrated speech and data traffic, 54–57, 75–77
interface, 47
International Alphabet No. 2 (IA2), 41–42
International Alphabet No. 5 (IA5), 37, 42
international standards, 18, 47
International Standards Organisation (ISO), 47
International Telecommunications Union (ITU), 18
'I' series Recommendations of (CCITT), 47

key system (PBX), 63

leased private circuit, 25–26, 58–60, 93–94, 116, 122, 127, 140, 144, 146, 153, 157, 160
limited availability, 90
line (termination) circuit, 62, 98–100, 103–104
link-by-link signalling, 26
list of requirements, 154–155
local area network (LAN), 67–74, 129–131, 147, 155–156
local circuit, 1–2, 14–15, 50–54, 58–59, 99, 115–116
local network, 14–15, 24, 50–54
local switching centre, 14–16, 27, 50, 54, 97–102, 110–111, 115–118
loop/ring-type local area network (LAN), 67–68
loss working, 6
lost traffic, 4

mathematical distribution, 89
mathematical model, 137
measurement (of performance), 136
measurement (of traffic), 4–5, 79–88, 103, 133
Mercury Communications Ltd., 31, 54, 120, 149, 151–153
mesh network, 64–65, 67, 109
message delay, 10–11
message handling time, 40, 85, 92, 128
message holding time or duration, 11, 84–86

message pair, 74, 85, 93
message switching, 40–42, 78, 84–86, 92, 103–105, 108–109, 127–128, 145–146
message transit time, 40, 92, 128, 146
microprocessor, 105–106
microwave radio transmission system, 23–24, 54
minimum cabling cost, 116, 120, 127
mobile radio network, 31–34, 152
mobile switching centre (cellular radio), 31–32
modem (data), 37, 157
modulation, 17, 19–20
modulation rate, 40–41
multi-loop data link/network, 66–67
multi-frequency (MF) signalling, 15, 18
multi-frequency (MF) telephone, 15
multiplexing: frequency-division (FDM), 17
    statistical (time-division), 45
    time-division (TDM), 19–22, 38
multi-point radio system, 54

net data throughput, 126, 129, 130, 145–146
network augmentation, 81, 85–86, 88, 91, 142–143, 146
network delay, 11–13, 43, 78, 84–86, 92–94, 125–129, 131, 145
network design and planning, 74–75, 94–95, 115–131, 135, 158–161
network terminating equipment (NTE), 52–54
network (traffic) management, 125
non-blocking, 99
non-coincident busy hours, 8

occupancy, 5, 84–85, 93–94, 122–123, 128–129, 146
octet (data), 37
offered traffic, 4
Open Systems Interconnection (OSI), 47–49, 74
    OSI Basic Reference Model, 48–49
operator's console (PBX), 61–63
optical fibre, 23, 54, 72–73
optical transmission (system), 23, 54–55
'out-of-area' local circuit, 115
overall grade of service (GOS), 91–92, 141–144
'own-switching-centre' connecting circuit, 98, 100–101

packet (data), 12–13, 37, 44–46, 85
packet assembler/disassembler (PAD), 44

packet delay, 13, 43, 86
packet handling time, 43, 86, 94
packet holding time, 13, 85–86
packet mode (data) terminal, 44
packets per second, 85–86, 94
packet switching, 42–46, 67, 76, 78, 85–87,
    93–94, 103–105, 108–109, 128–
    129, 146
parent (of switching centre), 2
partially-connected network, 109, 117–118,
    127, 144
performance capability, 132, 156
performance engineering, 132–135, 143
performance evaluation, 132–138
performance measurement, 136
performance targets, 139–147, 155
peripheral circuits (PBX), 61, 63
point-to-point (data) link, 65–66
point-to-point operation (wide area network),
    74, 131
point-to-point radio system, 54
polling mechanism, 67
'pool' signalling circuit, 98, 101–102
power failure, 139–140
predictive evaluation (of performance), 136–
    138
present value/present-value factor, 160, 162
private branch exchange (PBX), 2, 54, 58,
    60–65, 76–77, 82–83, 96–102, 115–
    118, 141, 144, 152, 155–156
private network: general, 8, 58–60, 75–77, 88,
    95, 139–140, 153, 154–161
    data, 47, 65–77, 85–87, 93–
    94, 125–127, 129–131, 145–
    147
    telephone, 60–65, 75–77, 79,
    81–83, 92, 116–118, 120–
    123, 140–144
probability, 89, 124, 126, 128, 131
processor control (of switching centre or
    PBX), 61, 102–109
processor load control, 106–109
protocol, 46–47, 68–71, 74, 130
protocol conversion, 48, 75
public network tariffs, 7, 140, 147–153
pulse-code modulation (PCM), 19–20, 23
    PCM (transmission) system, 19–22, 54

regeneration (pulse), 19, 21
registration (cellular radio), 32
reliability, 92, 97, 105, 156
repeat attempt, 6
repeater, 19, 23, 67–69, 73

response time, 86, 126, 129–130, 145–147
retry, 6
ring/loop-type local area network (LAN),
    67–68
router, 74, 131
route (traffic), 2–3, 7, 15, 79–81, 89–90, 98,
    101, 113, 117–121, 123–124, 141–
    144, 149
routing (of traffic), 46, 86, 92, 109–113, 120–
    124, 127, 129, 140, 142–143

satellite communications link/system, 58, 157
semi-permanently switched connection, 59–
    60
service-independent network, 56
set-up, 2–6, 26–27, 35, 52
set-up delay, 6, 26–27, 35, 52
signalling: general, 14, 39, 101–102, 103
    channel-associated, 26, 101
    common-channel, 27–30, 65, 76,
    101
    direct current, 15, 17–18
    end-to-end, 26
    link-by-link, 26
    multi-frequency (MF), 15, 18
    PCM system, 21–22, 101
    voice-frequency (VF), 18, 41
simplex (data) circuit, 10
simulation model, 133, 137–138
single-choice routing, 110–111, 120, 143
sizing, 93, 132–135
slotted-ring local area network (LAN), 70,
    130
sound waves, 14
space-division switching, 26
speech-transmission frequency range, 14
star-connected ring (LAN), 68
star network, 64–65, 67, 72
start-stop transmission, 36, 41
station (data network), 65, 68–69, 125, 130
statistical (time-division) multiplexing, 45
storage capacity, 40, 43, 84, 86, 92, 104, 108
storage element, 104
store-and-forward working, 10–12, 40–41, 78,
    84–85, 92, 108, 113, 128, 145
stored-program control, 105, 127
sub-system (data switching system), 104–105,
    128, 129
successful call (attempt), 3–4, 79, 83, 122–
    123
switchblock, switching element or switch
    unit, 61–62, 98–99, 104
switching: circuit, 35, 40, 55, 57, 75–78

four-wire, 26–28
message, 40–42, 78, 84–86, 92,
103–105, 108–109, 127–128, 145–
146
packet, 42–46, 67, 76, 78, 85–87,
93–94, 103–105, 108–109, 128–
129, 146
space-division, 26
tandem, 65, 99, 109–110, 117–
118, 120–121
time-division, 26–30
two-wire, 26–27
variable bit-rate, 56, 76, 114
switching centre: general, 1–2, 6–8, 41, 43,
46, 81, 84, 141, 143–144
local, 14–16, 27, 50, 54, 97–
102, 110–111, 115–118
trunk/toll, 15, 99, 110–111,
118–120, 124
switching system: general, 96–97, 113–114
data, 103–105, 108–109,
113
telephone, 97–105, 106–
108, 109–113
synchronous transmission, 37, 42

tandem switching, 65, 99, 109–110, 117–118,
120–121
target: average message-transit time, 146
average response time, 145, 147
overall grade of service, 141–144
tariffs (public-network), 7, 140, 147, 153
Telecommunications Act, 1984, 151
telegraphy, 9–10
telephone call, 1–6
telephone switching system, 97–103, 106–
108, 109–113
telephone traffic, 1–9
teleprinter, 10, 40–41
teletex, 42, 54, 152
teletraffic theory, 132–133
telex, 10–11, 39–42, 54, 78, 84–85, 92, 127–
128, 152
terminal (data), 9, 46–47, 125–126, 129
throughput (data), 86, 93–94, 104, 108
throughput efficiency, 126, 129, 130, 145–146
tie-lines (inter-PBX), 64, 116
time-division multiplexing (TDM), 19–22, 23,
38
TDM transmission system, 22
time-division switching, 26–30
token-access or token-passing data-transfer
techniques, 71–72, 130

token-bus or token-ring local area network
(LAN), 71–72, 130
toll switching centre, 15, 99, 110–111, 118–
120, 124
traffic balance, 141
traffic-carrying efficiency/performance, 5,
118, 120–121, 128, 132, 139–147,
156
traffic-carrying performance targets, 139–
147, 155
traffic concentration point or traffic concentrator, 2, 118
traffic engineering, 132
traffic flow: average, 4, 79–80
call-based, 1–9, 79–83
data 9–13, 78, 83–87
message-based, 10–12, 78, 84–85
packet-based, 12–13, 78, 83–84,
85–87
variability of, 7–8, 123
traffic forecasting, 79–88, 118–119, 133, 141
traffic-handling performance, 132–138
traffic measurement, 4–5, 79–88, 103, 133
traffic route, 2–3, 7, 15, 79–81, 89–90, 98,
101, 113, 115, 117–121, 123–134,
141–144, 149
traffic-route (termination) circuit, 101
traffic routing, 46, 86, 92, 109–113, 120–124,
127, 129, 140, 142–143
traffic table, 90
transceiver, 69
transmission: analogue, 14, 35, 58, 65
digital, 19–23, 26–29, 35, 38–
39, 45, 50–56, 48–60, 65, 76,
119
transmission loss, 15, 17, 18, 26
transmission network, 24–25, 39
transmission speed, 36–38, 42, 45, 52, 55–56,
58, 71–72, 93–94, 126–127, 129–
130, 145, 147, 153, 157–158
transmission system: FDM, 17
microwave-radio, 23–
24, 54
optical, 23, 54–55
PCM, 19–22, 54
TDM, 22
transponder, 73
tree network, 69, 72
trunk/toll switching centre, 15, 99, 110–111,
118–120, 124
two-wire circuit, 14, 16–17, 26–27
two-wire switching, 26–27

unidirectional circuit, 3
unsuccessful call (attempt), 4, 79, 83, 122–
123
unswitched data network, 74, 85, 93, 125–
127, 145

variability of traffic flow, 7–8, 123
variable bit-rate switching, 56, 76, 114
video, 24, 38–39, 55

virtual call working, 43–44, 113, 129
virtual circuit, 44
voice-frequency (VF) signalling, 18, 41
'V' series Recommendations (of CCITT), 47

wide area network (WAN), 74, 131, 147
wideband circuit, 38, 60

'X' series Recommendations (of CCITT), 47